T0185642

Standards and Innovations in Information Technology and Communications

Dina Šimunić • Ivica Pavić

Standards and Innovations in Information Technology and Communications

Dina Šimunić
Faculty of Electrical Engineering
and Computing
University of Zagreb
Zagreb, Croatia

Ivica Pavić
Faculty of Electrical Engineering
and Computing
University of Zagreb
Zagreb, Croatia

ISBN 978-3-030-44419-8 ISBN 978-3-030-44417-4 (eBook)
https://doi.org/10.1007/978-3-030-44417-4

This Springer imprint is published by the registered company Springer Nature Switzerland AG
The registered company address is: Gewerbestrasse 11, 6330 Cham, Switzerland

Preface

This book is important, because we believe that the reader will get better insight into the processes of standardization and innovation, as well as in their interaction and mutual interdependence.

We have been teaching the subject on technical standardization for more than 10 years at the University of Zagreb, Faculty of Electrical Engineering and Computing, Croatia. We received positive feedbacks from our former students once they became young engineers. This is the reason why we are so enthusiastic about this book, hoping that it will shine more light to the topic of standardization and innovation for the reader.

Thank you so much for reading!

Zagreb, Croatia

Dina Šimunić
Ivica Pavić

About the Book

This book is written in eight chapters, discussing various aspects of innovations and standards.

Chapter 1 is the introduction.

Chapter 2 gives an overview of innovation and standards geography, discussing innovation types and pillars and geographical connection to standards. It also shows sustainability of an ecosystem consisting of the technical standardization process, technical standards, and technical deployment of the overall country progress and of the innovation.

Chapter 3 discusses innovation and standardization stakeholders, starting with the four stakeholders in the innovation process. It also discusses introduction of the fifth stakeholder in the innovation process that avoids efforts in relation to the crucial standardization deliverable and prevents the so-called island solutions that are always an issue for the users.

Chapter 4 gives an overview of innovation and technical standardization documents with explanation of their mutual support and interactions.

Chapter 5 discusses innovation and technical standards life cycles and their organizations and perpetual interconnections in the time domain: when one process is coming to its end, the other process starts.

Chapter 6 introduces global innovation and standards circles or as they are named here global innovation circle (GIC) and global standard circle (GSC), which are discussed with all their elements and configurations. It also discusses three biggest and most recognized global standardization organizations: International Organization for Standardization (ISO), International Electrotechnical Commission (IEC), and International Telecommunication Union (ITU).

Chapter 7 presents further overview of supranational innovation and standards circles with their main stakeholders and configurations. It also discusses and presents, in relation to the supranational standards bodies on the European level, three European Standards Organizations (ESOs), i.e., European Committee for Standardization (CEN), European Committee for Electrotechnical Standardization (CENELEC), and European Telecommunications Standards Institute (ETSI).

Chapter 8 gives insight into usual and possible national innovation and standards circles with the stakeholders and the most important configuration types.

The book finishes with this chapter, hoping that the reader will get a flavor and clear idea into the complexity and interactions of the world of innovations and standards.

Contents

Chapter 1
Introduction

1.1 Introduction to the Innovations and Standards in the ICT Domain

It is quite incredible how the human society is, based on the technology, developing in the last 150 years. As a matter of fact, it corresponds to the development of the technical standardization in general. Therefore, it is not complex to conclude that the technical progress, which is based on the technical innovations, has the solid foundations and pillars in the technical standardization. This is an especial case for the development of the Information and Communications Technology (ICT) that finally brought the human society to a completely unpredictable (we can say: almost science fiction) future in a very short time period of cca 50 years.

Throughout this book, we will try to discover the connection, if any, between the innovation process and the technical standardization process.

Technical standards are voluntary documents, which help facilitate the trade between countries, create new markets, cut compliance costs, and support the development of national, supranational, and global markets. They enable much faster development of the innovations in the ICT sector, due to the spreading and interacting of the technical knowledge on the global level. Exactly the globally developed and accepted standards enabled a fast development and interoperability of ICT innovations and technologies steered by it. The winning case was the development of Global System for Mobile Communications (GSM) [1]. GSM is the second generation digital cell phone technology that was the first effort of mobile technology environment to create a transnational mobile system with the idea of seamless communications anywhere and anytime. GSM is developed as a set of ETSI specifications, becoming "de facto standard," that will be later explained in Chap. 3. This fact enabled GSM to spread all over the globe, and furthermore, it contributed to the global interoperability. After GSM, the world tried to develop the third-generation system (3G) of mobile telephony: Universal Mobile Telecommunications System (UMTS) that was supposed to enable user to travel around the globe with only one

D. Šimunić, I. Pavić, *Standards and Innovations in Information Technology and Communications*, https://doi.org/10.1007/978-3-030-44417-4_1

cell phone and without experiencing any kind of poor communication. However, this was not successful till now (even with the enrollment of 4G), but the world is getting closer and closer there. The next chance is 5G, where big changes of the society will appear in the relation to the society organization. 5G communications will enable platforms for vertical development of the society that never existed before. Therefore, we can expect frog leap and explosion of new approaches related to the concept: anything, anywhere, and anytime and in unlimited amounts. Autonomous traffic, energy grid, factories of the future, smart agriculture, and quality of human life, including personal health monitoring and treatment as well as environmental health monitoring, are only the most promising near-future use cases of 5G deployment in the smart city. Complexity of the smart city cannot be overcome without enough computational power in terms of artificial intelligence that is organized via wireless and wired communications, i.e., mighty variant of ICT. The 5G requires definition of clearly developed interfaces between 5G and all the older technologies than 5G (starting from 2G to 4G). All the mentioned goals are achievable only by global standards development that will be valid and applicable all over the globe.

The next successful global ICT story is the one of the Bluetooth [2]. Bluetooth was created and developed by few individuals within company Ericsson, with the simple idea for wireless connection of computers with printers. Instead of staying with the proprietary standard, the inventors decide to go for a global standard [3]. Nowadays, Bluetooth is applied for all kinds of connections, and it develops continuously all over the Earth.

Only the two presented examples can already indicate that there is a defined connection between standards and innovations. Standards are firing the innovations – innovations are integrated into standards. This is a perpetual life cycle that we tried to explain in the book, especially for the domain of information and communication technologies.

Bibliography

1. Global System for Mobile Communications, GSM, 3rd Generation Partnership Project, Technical Specification Group Radio Access Network, GSM/EDGE Physical layer on the radio path, General description, (Release 15), 2018
2. IEEE 802.15.1-2005, IEEE Standard for Information technology, Local and metropolitan area networks, Specific requirements, Part 15.1a: Wireless Medium Access Control (MAC) and Physical Layer (PHY) specifications for Wireless Personal Area Networks (WPAN), 2005
3. "Standardization and Innovation", Speech 2.1, Bringing radical innovations to the marketplace, Lars Montelius, Director General, International Iberian Nanotechnology Laboratory (INL), Professor of Nanotechnology at the Sweden Nanometer Structure Consortium at Lund University, ISO innovation, ISO-CERN conference proceedings, 13–14 Nov 2014

Chapter 2
Innovation and Standards Geography

2.1 Perpetual Interaction of Standards and Innovations

Standardization is known as a well-defined process that produces excellently organized documents the best possible organized documents at the current state of knowledge. It is a process that never stops. It is closely coupled with innovation and industry. However, perception of a "boring standardization" is present in our everyday life. Most probably, it is because very few people on this planet speak so-called standard language. It also means that most of the population in any country is defined as a nonspeaker of "standard language." Nonspeakers are most probably not aware of the importance of standardization processes and their role in the society, especially of their role in the innovation process.

A second kind of spoken language present in the research community may be called "research and innovation." Currently, this language is not understood by all the speakers of the "standard" language. The speakers of the two languages cannot understand each other. This is the factual case, because only a very small part of inhabitants of any of the two planets visit the other planet. It is also a perception of many that innovation and standard never meet: the two processes develop in entirely different directions (Fig. 2.1).

The inhabitants of the "research and innovation" planet live in a dynamic and exciting environment, usually without too many rules and restrictions. Innovation is defined as a "new idea, creative thought or new imaginations in form of device or method" [1]. This means that the innovation is a process of implementation of a new idea or an invention (as a creation of a new idea or a method) in a product, service, or process.

Technical innovation is, furthermore, related to a process of implementation of a new idea or an invention in a technical product, service, and process with a result in lower production costs and/or greater value added, as, e.g., higher product quality or new features. One may discuss how many forms innovation types can take. The "classical" innovation division can be set up according to the geography: it can be

D. Šimunić, I. Pavić, *Standards and Innovations in Information Technology and Communications*, https://doi.org/10.1007/978-3-030-44417-4_2

Fig. 2.1 Perception of the interaction between standards and innovations. Current perception shows no interaction at all between standards and innovations

on individual, company, governmental, national, regional, and international level. However, it is more usual to speak of four basic types of innovations:

1. Sustaining or incremental
2. Disruptive or creative destruction
3. Technology transferable
4. Transformational or radical

The sustaining or incremental innovation is probably the most common innovation type. It is used mostly in large companies. The sustaining innovation is based on the evolutionary changes of the products and markets to the best of the existing customers' satisfaction.

The disruptive or "creative destruction" [2] creates new markets. It is most common to a start-up mirrored in the "creative destructive" or disruptive shifts of technology, regulatory changes, or customers.

The technology transfer-based innovation means that the certain existing technology is applied in a technical, economic, or societal sector. Suddenly, an idea may come to apply this technology in some other sector or for some other application. If this new application brings much better results than the previous applications, we talk about technology transfer-based innovation.

The transformational or radical innovation opens completely new horizons and transforms paradigms in the technology (new industries), economy, or society.

All the four innovation types are shown on Fig. 2.2. The first innovation type, sustainable, is denoted as 1. S, the second, disruptive as 2.D, the third, technology transfer as 3.TT, and the fourth type, transformational innovation type as 4.T. They are placed in the society-technology plane. The abscissa represents society well-being value, or how large is the contribution of the innovation to the society well-being. The ordinate shows technology innovation level. For example, the most common innovation type, sustainable (1.S), brings certain technology innovation level and certain value to the society well-being, but the highest contribution in the whole plain is brought by the transformational innovation type (4.T). The disruptive innovation type has arrows, meaning that its society well-being value varies, dependent on the innovation itself, from giving the contribution as the sustainable innovation type (1.S) toward the technology transfer innovation type (3.TT) to the transformational innovation (4.T).

Usually, the free-spirited and innovative people tend not to bother about an existence of any kind of rules, which they perceive as being quite boring and even

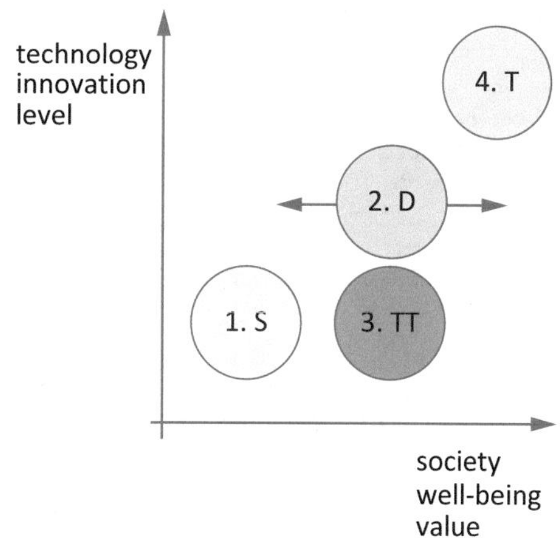

Fig. 2.2 Four innovation types in the society-technology plane. The sustaining innovation is denoted 1.S; the disruptive is denoted 2.D; the technology-transfer is denoted 3.TT; the transformational is denoted 4.T

"killing" creation. The other reason is that sometimes the innovative and creative people are introduced to rather old existing standards, being locked to the legacy systems. Thus, they conclude that the population of the "standard" planet generates the results that are of no interest for them. Furthermore, they withdraw from any interaction with the inhabitants of the non-exciting "standard" planet and also of their results. Another possibility is that if an average innovator bumps into the very new standards, she/he is so much challenged in terms of a possible not-yet-existing infrastructure that she/he loses interest in creating a new product according to the very new standard. Therefore, albeit without knowledge about it, the "standard" planet is defined as "boring," and, according to them, it should not be visited too often. Although the innovators are very creative, and, on the top, quite knowledgeable persons, the interaction and, what is even worse, the communication between the "innovator" and "standard" planets stop there.

In the other world of the "standard" planet, inhabitants use its own separate and completely different language and logic. On this planet, inhabitants hardly understand the life on other, much less organized, planets, especially on the "innovation" planet without the strict administrative rules.

Therefore, even though it is self-evident that the world of innovations and the world of standards makes a perfect match for an eternal happy life, the two worlds seldom talk to each other. Most probably, it is the consequence of two different cultures speaking in two different languages on two different planets, as depicted in Fig. 2.3.

Therefore, we can compare the innovation and standard planets to the stars. One planet can be the Sun, the other one the Moon. Most of us on the planet Earth are oriented toward the Sun. However, it is evident that the life on the Earth is not sustainable without a continuous change of the Sun and the Moon.

It is a fact presented later in this book that the first technological, and especially, electricity progress started exactly with the creation of standardization activities and bodies. To be even more precise, in the case of the electricity progress, the first two international standardization organizations that were established more than 150

Fig. 2.3 The planets of innovations and standards are like the Sun and the Moon for the Earth, being the progress of the human society

and 100 years ago still exist (International Telecommunication Union (ITU) [3]; International Electrotechnical Commission (IEC) [4]).

Everyone involved in any kind of technology standardization knows that there is a lot of knowledge and global coordination involved in each and every of the published standards, being on the global (international), transnational, or national level (Chaps. 6, 7, and 8 of the book, respectively). The most important characteristic of the standards is that they are, in principle, not a subject to intellectual property rights. This allows anybody, either in a private arrangement or in a small (or big) company to buy the standard and its contained knowledge for a relatively low price. Low price of a standard is enabled by the voluntary work of all the experts involved in its writing, as well as by the fact that the work is done by the nonprofit organizations (e.g., in the EU these are the European Committee for Standardization (CEN) [5] and the European Committee for Electrotechnical Standardization (CENELEC) [6] and the national standardizations organizations). The operational way of the standardization organizations ensures building in of all the existing technology knowledge from all the experts, independently of whether they are coming from small or big companies, from universities, or it they are independent researchers. The most significant property of the standard is that it is built by consensus, not only within the committee that develops the standard but also within various committees. For example, CEN has the agreement with International Organization for Standardization (ISO) [7] (so-called Vienna Agreement, 1991), CENELEC with the International Electrotechnical Commission (IEC) [4] (so-called Dresden Agreement, 1996, and Frankfurt Agreement, 2016), and the European Telecommunications Standards Institute (ETSI) [8] with the International Telecommunication Union (ITU) [3] (Memorandum of Understanding, MoU, 2000 and 2012). This means that all the relevant European Standards Organizations (ESOs) with a mandate from the European Commission (only three CEN, CENELEC, and ETSI) are coordinated with the global standardization. Furthermore, this means that ESOs are completely harmonized with the actions in the global standardization arena. This coordination saves a lot of effort and unnecessary additional work and time. On the other hand, it also ensures that any European standard contains global knowledge and is of global importance. Therefore, anybody who fetches any European standard is completely sure that it contains up-to-date global knowledge and that it is completely compatible and compliant on the global level.

The next factor contributing to the enormously important relationship between innovation and standards is a high dependence of the country's overall and,

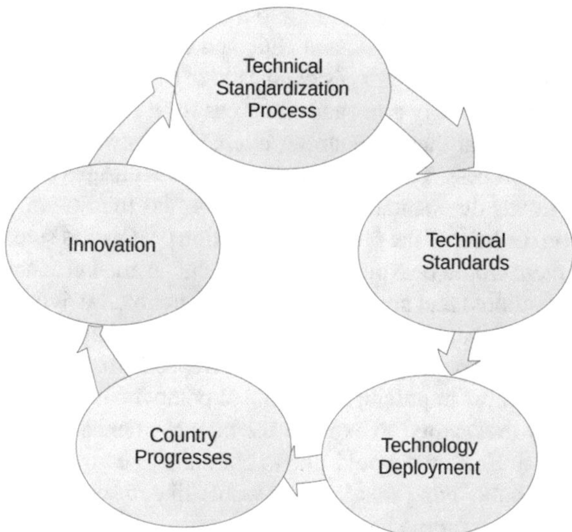

Fig. 2.4 Sustainable ecosystem consisting of technical standardization process, technical standards, technology deployment, overall country progress, and innovation

specifically, technological progress on the total knowledge in the country. One can specify a direct linear relationship between the overall state-of-the-art knowledge and its application in the country and the overall country progress [9].

Figure 2.4 depicts the whole closed self-reproductive ecosystem of the technical standardization process, technical standards, technology deployment, overall country progress, and innovation. The visible clear link between technical standards and overall country progress results in the innovation that again starts the process of technical standardization and technology deployment without an end. If this is applied to the whole region, or to the whole international arena, it means that the innovation takes place overall.

Thus, it can be easily concluded that use of standards is a "cheap-and-easy" way toward economic development of the country. This is since the variety of products can be introduced to the market. Also, the standardization process enables much quicker and broader (the speed and degree) adoption of new products and services. This places all national and, especially, international standardization organizations at the very specific location that enlightens continuous economic growth simply by taking care of sustainable dissemination of all the standards and especially of all the emerging technological standards on the national and international level. European Union is a special case related to the international scene, since it encompasses 27 different European countries, which follow the same legislation and standardization processes. It means that all the European countries follow also the same standards. Only if a certain national standard is of no interest to other European countries, it stays only as a national standard. Otherwise, it is in a definite timescale converted or transformed to a European standard. European standards are being developed in harmony with global standards (according to the previously mentioned agreements between CEN and ISO, CENELEC and IEC, and ETSI and ITU). Thus, it can be concluded that once when a country is connected to the global standardization processes, it has all the necessary tools and resources for a strong economic growth.

Our conclusion is that there is a strong interaction between the standardization and the innovation process. The standardization process ensures and shapes innovation pattern. Without the standardization process, the innovation process would either develop on the base of the previous innovation or it would spread too wide. In the latter case, there would be a plethora of products (as final products of a plethora of technical innovations) that are completely incompatible. As it is known from the past, incompatibility leads to the very local production and usage, meaning that the use of the same and compatible products outside a specific region, especially related to ICT, would be almost impossible. The global component of the development of the human society that started to explode the human science in an unseen dimensions would be lost. Since we would like to continue to enjoy the increasing comfort, healthcare, and the living standard, we would like to support the growth of the standardization process for the benefit of all.

Now we can redraw Fig. 2.1 from the beginning of the chapter. It is shown on Fig. 2.5.

According to [10], standards provide as much knowledge as imported knowledge from licenses and as half of knowledge deriving from patents.

The *Oslo Manual* [11] defines innovation as a new idea or an invention and its implementation. The diffusion and overall application of innovation in the society define social and economic impact of the innovation. Thus, the innovation can have a societal, a technical, or an economic nature, or any combination of all three. Furthermore, this proves once more that the innovation impacts productivity, economic growth, and, thus, the general well-being. At societal level, innovation serves to satisfy current and future human needs. At individual and company levels, innovation serves to increase market shares or general profits. Therefore, it is quite important to measure the extent of the innovation results on the societal and private levels. So, what are the innovation outcomes? They could be productivity, profits, jobs, higher well-being, or better environmental care. But the real question is how to measure innovation results, when they can be distributed over longer period of time, over many organizations and individuals, over larger geographical area, etc.

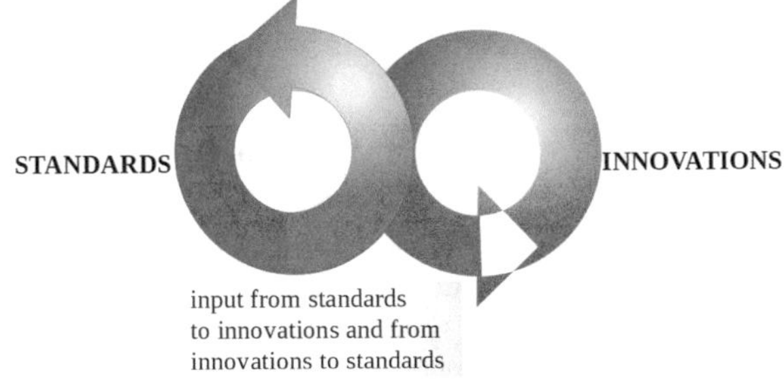

Fig. 2.5 Actual relationship between standards and innovations: The clear intersection area of the interaction between innovations and standards in the both directions

United Nations' (UN) System of National Accounts (SNA) [12] provides a globally adopted framework for measuring the economic activities of production, consumption, and accumulation. SNA defines concepts of income and wealth. This framework permits an integration of innovation data with other SNA consistent statistical sources. Therefore, SNA is considered as the best measuring innovation platform, and its terminology should be consistently followed.

SNA defines four major relevant economic sectors for an innovation, relevant for performing innovation measurements in institutional units with the legal responsibility of their actions:

- Business enterprise [13]/corporate sectors are corporations/business entities whose main products are physical products, processes, and services.
- General government units redistribute income and wealth with the nonprofit institutions included. Their main products are physical products and services, usually on a nonmarket basis.
- Households units are in reality one or more individuals. Their role is to provide labor and consumption of good and services and to produce market products, processes, and services, as entrepreneurs.
- Non-Profit Institutions Serving Households (NPISHs) do not produce market services, because their main mode of operation is voluntary. NPISHs can be a part of a business enterprise or, equally well, of a general government sector.

It must be noted that in SNA the product can be a good or a service. It results from production activities for a final consumption, for further exchange or as a beginning of the new production or for investment. Goods are objects with established ownership rights (that are easily transferable from one owner to another) and with existing current or potential demand. Service affects user conditions or facilitates the exchange of products (e.g., in financial sector). Examples of changes of user conditions are changes in the condition of users goods (e.g., materials from one company are transported to another company which makes a certain product), in the physical condition of a person (e.g., producer provides accommodation, transfer, medical or physiotherapy treatments), and in the psychological condition of a person (e.g., producer provides education, advice, and entertainment service; these are often digitally delivered services).

One of the logical measures is an innovation project. It is a logical measurable unit in large companies, as a result of internal policies, processes, and procedures (examples are research and development engineering, design and other creative activities; marketing and brand equity activities, Intellectual Property (IP)-related activities, employee training activities, software development and database activities, activities related to the acquisition or lease of tangible assets, and innovation management activities). However, the situation is not the same for a small start-up company that invests full resources to a single innovation, not even considering calling it an innovation project.

The innovation measurement can be performed at the household level, because of the recently developed required changes in terms of users' behavior in recycling and sustainability. The measurement can also be performed in a certain geographic area for data collection with defined measurement strategies.

However, it seems that there is no simple relationship between innovation and macroeconomic effects on industries, markets, or economy. The first who introduced the concept of disruption, as relevant factor in the relationship between innovation and economy, was Schumpeter [2]. As mentioned earlier, the concept was called a "creative destruction." The term describes the phenomenon quite well. This is the reason for the later use in the "transposed" version, simply as the "disruption." Disruptive processes give possibility of finding new ways of production or of finding completely new industries. Therefore, we can conclude that the long-term economic growth depends on the innovation processes, giving it the utmost priority to study them in order to ensure it. The OECD study of innovation in 1997 [14] resulted in understanding that the innovation is the result of complex interaction of various inputs and that it is not a linear and sequential process. Therefore, the only way of understanding interrelationship between all the inputs of innovation systems is to follow the long-term systematic and interdisciplinary approach. The entities from the four mentioned SNA economic sectors are innovation systems that are often related and interrelated to systems on the local, regional, and global scene. Therefore, all measurements are responsible to collect data not only on an entity level but also on the national, regional, and global level. Fortunately, nowadays is the collection enabled by using disruptive innovative tools and networks from Information and Communication Technology (ICT) industry. This is also why the current systematic perspective is very much developed, when compared to the earlier times without the global interconnection. It is even possible to coordinate global innovative systems to develop a healthier planet Earth with happier and healthier individuals [15]. As a matter of fact, the awareness of the latter brings the item of evaluation of innovation processes high on the world's agenda.

In short, the subject innovation is based on the four pillars, according to the *Oslo Manual* [11]:

- Knowledge (K)
- Novelty (N)
- Implementation (I)
- Value creation (V)

Figure 2.6 shows four pillars of innovation, with K denoting knowledge, N novelty, I implementation, and V value creation). It is important to understand that together they form the necessary condition for the innovation appearance. It is their interaction, their interplay, and, finally, their synergy that brings the innovation.

Fig. 2.6 Four pillars of innovation appearance are knowledge (K), novelty (N), implementation (I), and value creation (V)

Knowledge is the obvious, the most important, and the obligatory pillar for any innovation. An especial aspect of knowledge is that it can be gathered by own experimental and research work or by a knowledge transfer in a direct or indirect way. In any case, the knowledge is apprehended and gathered by cognitive processes, meaning that it requires initiation of certain learning processes in the recipient. Knowledge was very exclusive in the past, where only the certain layers of the societies had exclusive access to it. The increase of the global interconnection, enabled by the ICT systems, showed the need for a transparency of all aspects and layers of the modern complex societies, driving the further dynamic progress of the planet. Consequently, many societies concluded that the exclusive rights, like Intellectual Property Right (IPR) that previously served to protect from rivalry, presents a certain degree of inhibition of innovation in the field of technology. Not only that, it was concluded that the society and the economic growth could feel the consequences of the inhibition. Therefore, the current strategy of the European Commission is the promotion of an open society, founded on an open-source software and open scientific publications [16]. This policy motivation enables such a society to make wide available knowledge overflow for the creation of new knowledge and, thus, the new innovative and yet unknown social and economic values. However, the first and necessary condition for this is a digitalization of the society, of all the systems in the country and, e.g., in the European Union, opening all the new possibilities for the citizens. Only after proper digitalization, an information can be accessed and organized, opening the way to any entity of the society to be able to achieve the desired level of knowledge to generate an innovation. As stated earlier in this chapter, standards contain a lot of knowledge as a result of internationally driven and coordinated consensus. Thus, we are free to conclude that the standards with their inherent knowledge represent the main pillar of the innovation.

Novelty of a product, process, or service is characterized by its use, primarily by the comparison with already existing products, processes, or services on the market. Novelty can be assessed or measured either in terms of physical technical measures, like material properties (e.g., weight), electrical properties (e.g., dissipated power, heating), or communication properties (e.g., velocity of communication, throughput), or in terms of user experience measures, like usability of the service, etc. The user experience measures are much more difficult for a cold and objective data gathering and their organization, since different users respect more of some than of other qualities (e.g., easy use in relation to data processing).

Implementation is the very next important pillar of innovation, because the definition of an innovation is related to it. If there is no implementation, an innovation is not the innovation, but it remains an idea, an invention, or maybe a prototype. Therefore, an implementation must be carefully planned after a clear definition and understanding of the innovation's novelty. Implementation usually requires a lot of work related to understanding user's needs and wishes, as well as current technology level for all the aspects of the innovation, being a product, a process, or a service. Finally, it means that the group of experts dealing with the implementation needs to take care of following it and redesigning it when necessary.

Value creation is the fourth pillar, related to the economic or social growth. It can be assessed only after some period of implementation, because various stakeholders can benefit with various values. The value creation, especially for government policy makers, is not only about economic profitability, but rather as a measure of the sustainable society progress in total.

In the modern world, it is crucial to understand the potential of digitalization, as a means not only for the business transformation but also for the society and the economy in general.

The digital service has the unique property of a rapid obsolescence and a replacement that was never present before to such an extent in the human society. Exactly this property is the basis for an incredible growth of innovation, connecting the whole globe to a huge single innovation process, and to the growth of transparency, happiness, and economic growth, leading to an overall well-being. The innovation processes are initiated for digitalization itself but also as a key driver for all other economic and societal growth.

Examples of such innovation processes in digitalization are:

- Information gathering, a core business of companies selling and developing information content
- Data development activities, a core business for companies working on improvement of business decision process in relation to innovations
- Data management knowledge, a core business of companies performing surveys and trying to assess innovation probability based on the data management and advanced technology development and use
- Knowledge flows, a core business of companies that analyze innovation process, relevant to digitized world and its decentralized collaboration model
- External factors, a core business of companies studying influence of, e.g., digital platforms and users' trust used by other entities, in order to produce another innovation-based product
- Measurement activities, a core business of companies collecting information on an innovation outside business sector
- Measurement activities for a visualization and analysis of innovation activities
- Measurement activities of identifying the key business/innovation partner
- Measurement activities of implementing more secure and easier electronic collecting methods
- Measurement activities of using statistical data on innovation and business

In order to be able to perform all the measurement activities in innovation by means of digital technologies, it is necessary to define user needs and to understand drivers and factors of innovation in various countries, communities, industries, and individuals. According to the *Oslo Manual* [11], three stakeholders are very much interested in usage of innovation data: academics, managers, and policy makers or policy analysts.

Bibliography

1. Merriam-Webster, Internet page: www.merriam-webster.com. Retrieved 27.8.2019
2. J.A. Schumpeter, *The Theory of Economic Development: An Inquiry into Profits, Capital, Credit, Interest and the Business Cycle*, translated from the German by Redvers Opie, (2008) (Transaction Publishers, New Brunswick/London, 1934)
3. ITU, Internet page: www.itu.int. Retrieved 17.8 2019
4. IEC, Internet page: www.iec.ch. Retrieved 20.8.2019
5. CEN, Internet page: www.cen.eu. Retrieved 20.8.2019
6. CENELEC, Internet page: www.cenelec.eu. Retrieved 19.8.2019
7. ISO, Internet page: www.iso.org. Retrieved 19.8.2019
8. ETSI, Internet page: www.etsi.org. Retrieved 18.8.2019
9. K. Blind, *The Impact of Standardization and Standards on Innovation* (Manchester Institute of Innovation Research, Manchester Business School, University of Manchester, Manchester, 2013)
10. K. Blind, R. Bekkers, Y. Dietrich, E. Iversen, B. Müller, T. Pohlmann, J. Verweijen (2011) EU Study on the Interplay between Standards and Intellectual Property Rights (IPR), 2011, commissioned by the DG Enterprise and Industry
11. OECD/Eurostat, *Oslo Manual 2018: Guidelines for Collecting, Reporting and Using Data on Innovation*, 4th edn. (The Measurement of Scientific, Technological and Innovation Activities, OECD Publishing, Paris/Eurostat, Luxembourg, 2018)
12. European Communities, International Monetary Fund, Organisation for Economic Co-operation and Development, United Nations and World Bank, *System of National Accounts* (European Commission, 2009)
13. OECD, *Frascati Manual 2015: Guidelines for Collecting and Reporting Data on Research and Experimental Development* (The Measurement of Scientific, Technological and Innovation Activities, OECD Publishing, Paris, 2015)
14. OECD/Eurostat, *Proposed Guidelines for Collecting and Interpreting Technological Innovation Data, Oslo Manual* (OECD Publishing, Paris, 1997)
15. G20 Innovation Report (2016) Report prepared for the G20 Science, Technology and Innovation Ministers Meeting, Beijing, China, 4 Nov 2016, OECD, Better policies for better lives, OECD Publishing
16. European Commission, *Policy Recommendations – Study, Cost-benefit Analysis for FAIR Research Data* (European Commission, 2018)
17. OECD, *The Innovation Imperative, Contributing to Productivity, Growth and Well-Being* (OECD Publishing, Paris, 2015)
18. OECD, *The OECD Innovation Strategy: Getting a Head Start on Tomorrow* (OECD Publishing, Paris, 2010)
19. OECD, *The Future of Productivity* (OECD Publishing, Paris/Eurostat, Luxembourg, 2015)

Chapter 3
Innovation and Standardization Stakeholders

3.1 Innovation Stakeholders

In general, there are four main stakeholders in the innovation process:

- Inventors
- Entrepreneurs
- Marketers
- Users

An innovation starts with inventors, without whose idea the innovation would not be possible. As an example, the inventor can be a lone inventor, or she/he can work in a more organized environment, e.g., in a company. The lone inventor is committed to the invention of a novel product. In most cases, the invention is based on an idea that could make a radical change in one of the basic activities of human life. Inventors may be scientists, artists, and/or any active societal player. They may or may not handle product design, marketing, and sales well.

The next most logical stakeholder is the entrepreneur. He brings the invention to the market, by financing the development, production, and diffusion of a product into the marketplace. Examples of the entrepreneur include industries, foundations, and governments.

Another stakeholder is the marketer, whose role encompasses creative, analytical, digital, commercial, and administrative responsibilities. The responsibilities of the marketer include overseeing and developing marketing campaigns, conducting research and analyzing data for an identification and a definition of an audience, presenting ideas and strategies and promotional activities, organizing events and product exhibitions, coordinating internal marketing events, monitoring whole performance, and managing campaigns on social media.

The fourth and the last, but not the least, stakeholder is the user. Without the user input, the cycle of innovation does not exist. The circle of innovation stakeholders is shown in Fig. 3.1.

D. Šimunić, I. Pavić, *Standards and Innovations in Information Technology and Communications*, https://doi.org/10.1007/978-3-030-44417-4_3

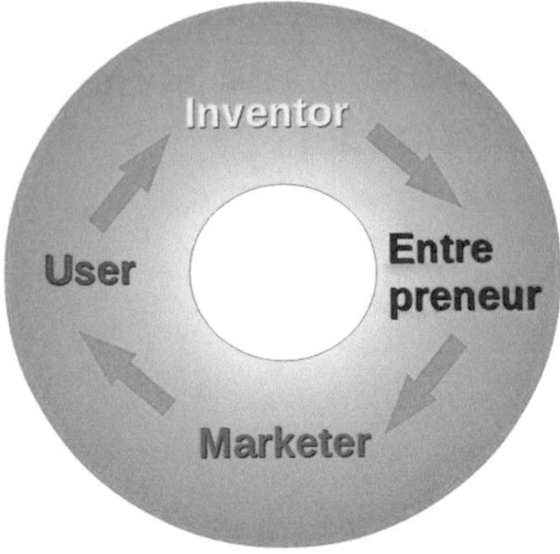

Fig. 3.1 The closed circle of innovation stakeholders' interaction

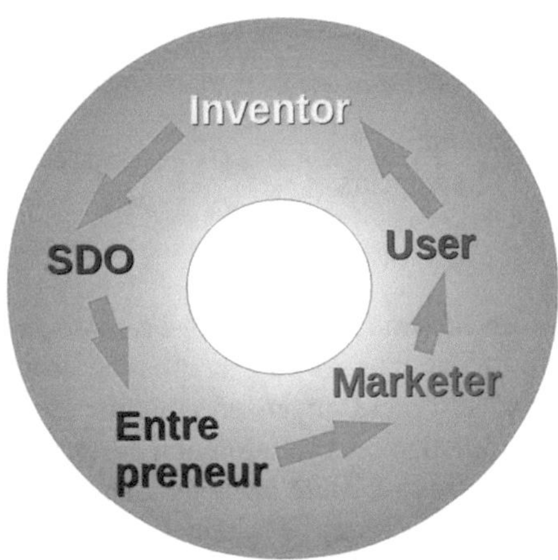

Fig. 3.2 Recently updated innovation stakeholders circuit with the four basic stakeholders from Fig. 3.1 and with the added fifth stakeholder: SDO

Any of the four stakeholders can be an intellectual property owner. In the recent history, Fig. 3.1 is expanded by another stakeholder: this is the Standards Developing Organization (SDO). SDO becomes an innovation stakeholder (Fig. 3.2) to avoid wasting duplicating effort on the elaboration of a standardization deliverable that could subsequently have serious problems with issuing or maybe that could be even blocked by the existing essential Intellectual Property Rights (IPR). Another reason for the SDO to become one of the innovation stakeholders is to enable another innovation stakeholder and very important societal stakeholder (a user) to overcome the so-called island solutions. "Island solutions" are based on a single technology, and it is impossible or almost impossible to combine them with any other already existing products. If the IPR is incorporated to the standard, users (consumers) will have all interoperable products in action from the start of the system operation.

Another set of stakeholders is involved in innovation measurement activities by means of digital technologies. The measurements are a necessity due to the required definition of user needs and due to the necessity to understand drivers and factors of innovation in various countries, communities, industries, and individuals. According to the *Oslo Manual* [1], there are three possible users of innovation data:

1. Academics
2. Managers
3. Policy makers/policy analysts

According to the same source, all three kinds of users would like to obtain comparable data across industries, regions, and time. They would like to keep up with changes of innovation (an example is an open innovation). Also, they would like to enable analyses of innovation impacts on innovative organizations or supranational and national economies. Finally, they would like to provide data on the factors enabling or hindering innovation as they would like to link innovation data to other data, such as individual users of innovations.

1. Academics would like to try to improve society's understanding of innovation and its socioeconomic effects. The main goals of the research are understanding and independent interpreting innovation outcomes, so that it can be an important contribution to innovation policy development. In addition, the researchers can provide hypotheses for new theories of innovation development that could contribute to the economic development and growth.
2. Managers can be the main drivers for innovation in the company, if they understand their role well. They should appoint persons/teams for the collection of the data on innovation, which then directly influence their own innovation capabilities. This is a sister process to one in ISO 9000 [2], i.e., an improvement of the management of the own company.
3. Policy makers, as well as policy analysts, have a tremendous interest in the innovation [3, 4] in all industries and SNA sectors [5]. Therefore, they require results of the research and the main indicators, as results of the measurement studies. All indicators have to be benchmarked, in order to achieve international harmonization on use in different national economies. Even though it is possible to benchmark indicators for a majority of cases, it is unfortunately not possible for all, due to contextual and, in general, cultural differences. However, digitalization contributes to the harmonization of defining new data collection or links to the existing publicly available sources. All the data are available due to the possibility of digital storage. Therefore, these can always be pulled from the storage, and together with newly gathered data, they may contribute to even higher reveal of the innovation process.

3.2 Standardization Stakeholders

International Organization for Standardization (ISO) [6] defines a standardization as "an activity of establishing, with regard to actual or potential problems, provisions for common and repeated use, aimed at the achievement of the optimum degree of order in a given context." Thus, processes of formulating, issuing, and implementing standards form the vital part of standardization. Standardization improves suitability of goods, processes, and services for their intended purposes. It also prevents barriers for trading, and it facilitates technological cooperation among different parties. Any topic that is being worked on in the standardization is called a standardization subject, and it is usually a good, a process, or a service. With the technological advancements, it is not possible to cover all aspects of a subject by only one standard. Therefore, standardization is usually limited to cover only particular aspects of a subject.

Standards work with the "state-of-the-art" knowledge, i.e., with the developed stage of technical capability of goods, processes, and services, based on the relevant consolidated findings of science, technology, and experience at a given time. The main point of relevance of a standard is that a lot of expertise, knowledge, and experience of various stakeholders of a good, process, or service have to be brought together. Nowadays, standards are not only technical documents that describe good, service, process, or their interoperability, but they are also essential for a global security and safety and higher quality of life. It is important to note that standards start to be more an alternative to formal regulation, meaning that they have become to be in the focus of interest to European policy makers and regulators. This is possible, because necessary conforming to standards can become a means of reducing actions by regulatory authorities, like enforcing inspections.

Standardization level refers to geographical, political, or economic extent of standardization involvement. Global standardization is a standardization in which involvement is open to relevant bodies from all countries, i.e., it is realized globally. In supranational standardization, an involvement is open to relevant bodies from countries from only one geographical, political, or economic area (region) of the world. When standardization takes place at the level of the specific country, it is called the national standardization.

This chapter will give an overview of a representative of supranational standardization stakeholders: the European Union.

In the standardization process, stakeholders can be standard bodies, public authorities, regulators, enforcement bodies, testing and inspection bodies, business and industry associations, professional bodies, producers, sellers, companies, trade associations, foundations, users or buyers, consumer organizations, environmental organizations, trade unions, or any other association representing them. They can be grouped into four main groups:

- Standards Body
- Authority
- Industry
- Societal stakeholders

Bodies responsible for standards and regulations have a central position in the standardization stakeholders circle (Fig. 3.3). They are legal or administrative entities with specific tasks and composition. For example, a standardizing body is a body dealing with standardization activities. Standards Body, SB, is a standardizing body for standards development, i.e., Standards Developing Organization, SDO, or

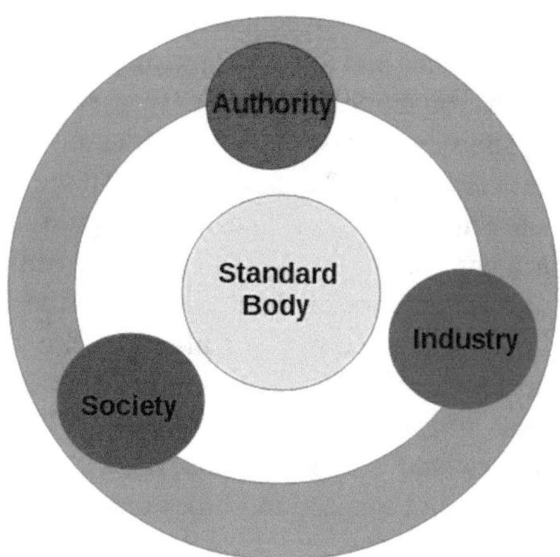

Fig. 3.3 Standardization stakeholders circle

standardization organization for standards setting, i.e., Standards Setting Organization, SSO. SSO is any organization with basic activities: development, coordination, public information, maintenance and correction, improvement, republishing, explanation and maintenance, or any work on technical standards with the intention of satisfaction of needs of certain groups. Therefore, a Standards Body is a recognized body on national, supranational, and international level with the main activities in accordance with its statute, such as preparation, approval, or accepting the standards, which are accessible by public.

An organization is a body that is based on the membership of other bodies or individuals with an established constitution and its own administration. In terms of standardization activities, standardization organizations can be grouped according to the role, position, and influence area. Namely, standardization organizations can be active on local, national, supranational, and global level. A National Standards Body, NSB, is a Standards Body recognized on national level. NSB has a right to be a national member of according international and Supranational Standardization Organizations. For example, in Germany, the National Standards Body is the German Institute for Standardization (*Deutsche Institut für Normung e.V., DIN*); in the Republic of Croatia, National Standards Body is the Croatian Institute for Standardization (*Hrvatski zavod za norme, HZN*). Supranational standards organization (SSO) is a standards organization with the open membership to the relevant national body from each country within the defined geographical, political, or economic region. An example of supranational region is the European Union. A global standards organization is a standards organization with the membership open to the relevant national body from any country.

Furthermore, the standardization organizations could be grouped also according to technology or economy criterion to organizations for standards development and organizations for standard setting. The further possibility of grouping could be according to the influence to governmental, quasi-governmental, and nongovernmental and, for example, nonprofit organizations. The grouping could be also performed according to the covered area as limited (e.g., in the local or global community or in government) and diffused (e.g., de facto standardization) or according to the application to voluntary or obligatory (e.g., regulatory activities) standardization.

The next important stakeholder in a standardization stakeholders circle is the authority (Fig. 3.3). It is a body with the legal powers and rights, acting on a supranational, national, or local level. Regulatory authority is responsible for an adoption or a preparation of regulations. Enforcement authority enforces regulations, and it may or may not be the regulatory authority. Regulatory standardization organizations, most often public authorities, are stakeholders in standardization process of the formal standards, especially in the approval phase. In the standards development, not only public authorities but also business and industry associations and professional bodies are important stakeholders giving considerable contribution. Testing and inspection bodies can be a part of the authority or of the next stakeholder, industry.

On Fig. 3.3, the third standardization stakeholder is the industry. All producers, sellers, companies, business and industry associations, foundations, professional bodies, trade associations, etc. belong to this category. The industry is a stakeholder with an economic and business interest: to finance the production and to get income out of the production and selling of the product that is a result of the standard.

Recent development of ICT products requires higher interaction of a user and a product. Thus, it became relevant to include societal stakeholders in the standardization process. Societal stakeholders aim to take care of priorities and concerns of the general population or certain stakeholders. They typically represent a wide cross section of society with different interests such as small and medium enterprises, consumer organizations, trade unions, and environmental organizations. In this way, they ensure relevance of developed goods and services to market expectations, e.g., with safeguarded safety and health of workers regarding environmental protection.

Users or consumers are usually represented by the consumer organizations as nongovernmental associations of a civil society. Standards users are global general populations, whose confidence in a certain product raise naturally with standards.

Trade unions are important societal stakeholders, since they have expertise especially in workers' health and safety challenges related to product safety, environment, service, energy, transport, or, e.g., nanotechnology.

Environmental organizations are important in relation to standardization of measurement methodologies to monitor environmental quality of human environments. Environmental quality is, for example, reflected in energy efficiency. This is especially the case in a development of a modern smart city, where ICT with related standardization plays a very crucial role.

An example of the transnational standardization stakeholders circle is the European Union. EU Regulation 1025/2012 [7] on European standardization requires facilitated access of societal stakeholders in the standardization process, in order to encourage representation of organizations dealing with social and environmental sustainability on national and European level. Consumers are represented via European Association for the Co-ordination of Consumer Representation in Standardization, known as European consumer voice in standardization, ANEC [8];

employees and workers by European Trade Union Confederation, ETUC [9]; and environment by European Environmental Citizens Organization for Standardization, ECOS [10].

Table 3.1 shows the list of Technical Committees (TCs) and Working Groups (WGs) in CENELEC with ANEC participation.

Table 3.1 shows the variety of TCs and WGs in which consumers are present via ANEC. This shows a big interest of consumers for shaping future products through their participation in the standards development. Shaping the European standards is necessary due to the requirements and needs for safe and accessible designed products or services before placing them to the market. CEN and CENELEC member countries ensure that European standards are becoming also national standards. In this way, European standards meet the needs of their consumers, i.e., of all citizens of European Union for their benefit and well-being. Technical experts involved in the standardization process benefit from consumer representatives' input on real-life consumer-oriented views. This benefit is particularly of a high appreciation when the European Commission gives a mandate to the European Standards Organizations (ESOs) to issue European Standard (ES), as a support of European legislative acts: directives or regulations for certain "sensitive" areas. One such example is toys, where safety requirements are of essential priority. This is also valid for safety requirements for other areas such as machinery, mobile phones, personal protective equipment, etc.

Trade Unions actively contribute to the standards development with their experience of workers' shop floor, demands and expectations, especially related to workers' health and safety challenges in a product safety, environment, service, energy, transport, or nanotechnology. European Trade Union Confederation (ETUC) [9] is a representative of European trade unions in services related to maintenance and facility management. In the standardization of industrial advanced manufacturing, ETUC deals with various topics, like Intelligent Transport Systems, bio-based products, and machine-to-machine communication.

In the field of machinery and ergonomics, European Trade Union Institute (ETUI) [11] is the ETUC independent research and training center. It represents European trade unions. ETUI insofar dealt with "mandated" standards supporting product Directives, with priority given to machinery and ergonomics standards. Standardization has the potential to provide the platform for a collaborative work between engineers, employers, workers, manufacturers, researchers, and governments who may contribute to the better health and safety through a consideration of design issues. Through standardization, trade unions can explore pathways to deliver the aim of putting workers' knowledge to its best use in the improvement of the working environment.

Table 3.2 shows the list of Technical Committees and Working Groups in CEN-CENELEC with ETUC participation. The variety of topics of the TCs and WGs shows the deep involvement of ETUC in the sEU standardization process.

Trade unions participate in the standardization due to many reasons. One of them is that the standardization nowadays affects a wide range of issues such as postal services, open-source software, sustainability, nanotechnologies, and management

Table 3.1 List of Technical Committees and Working Groups in CENELEC with ANEC participation

Name	Title
CLC/BT	CENELEC technical board
CLC/TC 34A	Lamps
CLC/TC 59X	Performance of household and similar electrical appliances
CLC/TC 59X/WG 01	Laundry appliances
CLC/TC 59X/WG 01-06	Washing machines
CLC/TC 59X/WG 01-07	Spinning efficiency and accuracy
CLC/TC 59X/WG 01-08	Rinsing performance
CLC/TC 59X/WG 01-09	Dryer
CLC/TC 59X/WG 01-10	Dosage of detergent
CLC/TC 59X/WG 01-11	Washer-dryers
CLC/TC 59X/WG 01-12	Commercial laundry machines
CLC/TC 59X/WG 02	Dishwashers
CLC/TC 59X/WG 06	Surface cleaning appliances
CLC/TC 59X/WG 06-01	Water filter vacuum cleaners
CLC/TC 59X/WG 06-02	Uncertainties for vacuum cleaners
CLC/TC 59X/WG 06-03	Commercial surface cleaning appliances
CLC/TC 59X/WG 06-04	Durability of suction hoses
CLC/TC 59X/WG 07	Smart household appliances
CLC/TC 61	Safety of household and similar electrical appliances
CLC/TC 61/WG 01	Relations between standardization and legislation
CLC/TC 61/WG 04	Use of appliances by vulnerable people, including children
CLC/TC 61/WG 06	Activities related to the new MD
CLC/TC 61/WG 07	Toys
CLC/TC 61/WG 08	Toy-like (child appealing) appliances
CLC/TC 62	Electrical equipment in medical practice
CLC/TC 62/WG 01	Medical beds for children
CLC/TC 100X	Audio, video, and multimedia systems and equipment and related sub-systems
CLC/TC 106X	Electromagnetic fields in the human environment
CLC/TC 108X	Safety of electronic equipment within the fields of audio/video, information technology and communication technology
CLC/TC 108X/WG 03	Sound pressure related to portable music players
CLC/TC 108X/WG 04	Investigation around the future of EN 41003 related to new hazard based standard
CLC/TC 111X	Environment
CLC/TC 116	Safety of motor-operated electric tools
CLC/TC 116/WG 05	Particular requirements for electric motor-operated lawn and garden machinery
CLC/TC 205	Home and building electronic systems (HBES)
CLC/TC 205/WG 16	Standards for intelligent home and building/smart houses
CLC/TC 205/WG 18	Smart grids

(continued)

Table 3.1 (continued)

Name	Title
CLC/TC 216	Gas detectors
CLC/TC 216/WG 09	EN 50291-1:201X
CEN/CLC/TC 10	Energy-related products – material efficiency aspects for ecodesign
CEN/CLC/TC 10/WG 2	Durability
CEN/CLC/TC 10/WG 3	Upgradability, ability to repair, facilitate re-use, use or re-used components
CEN/CLC/TC 10/WG 6	Documentation and/or marking regarding information relating to material efficiency of the product

systems. The management system requires a systematic approach to processes of an organization, e.g., how the work is organized. Furthermore, this means that standardization impacts the working conditions. This gives rights to trade unions for having a legitimate voice in the standardization process. Besides, standards serve as strategic industrial tools that nowadays modify competitive companies' position in the market. Finally, the European Commission requests the ESOs in relation to the Directives that are later on transposed to the national legislative acts, giving also a reason for the trade union involvement.

An individual trade union member from a small organization can experience difficulties for a participation in the complex world of European standards. Especial challenge could be a physical attendance at all the meetings. The complexity of the standardization process requires timely and accurate action, reflected by writing and following requirements and procedures, as well as by analyzing correspondence between experts. Since European standardization is an interplay process between activities on the national level and on the European level, the usual way for an individual is to access it via national trade unions complemented by European networking and coordination. Nowadays, workers' interests have to be represented in the standardization process, especially because of the required possible significant changes to health and safety of working conditions, enabling an anthropocentric approach in the human-machine interface.

National Standards Bodies allow every citizen to play an active role in developing standards. Examples can be given for CEN National Standards Body [12] and CENELEC National Committee [13].

ETUC national trade unions also provide help related to a support in the standardization activity. For example, EUROGIP [14] is a French organization that is involved in the experts' coordination in the development processes of workers' health and safety line of the European and International Standards. Another organization is active in Germany: KAN [15] ensures interests of social partners, German State, German Social Accident Insurance (DGUV) and DIN in standardization of occupational health and safety topics.

Table 3.2 List of Technical Committees and Working Groups in CEN-CENELEC with ETUC participation

Name	Title
CEN/BT	CEN technical board
CEN/TC 278	Intelligent transport systems
CEN/TC 278/WG 1	Electronic fee collection and access control (EFC)
CEN/TC 278/WG 3	Public transport (PT)
CEN/TC 278/WG 7	ITS spatial data
CEN/TC 278/WG 8	Road traffic data (RTD)
CEN/TC 278/WG 9	Dedicated short-range communication (DSRC)
CEN/TC 278/WG 10	Man-machine interfaces (MMI)
CEN/TC 278/WG 12	Automatic vehicle identification and automatic equipment identification (AVI/AEI)
CEN/TC 278/WG 13	Architecture and terminology
CEN/TC 278/WG 15	eSafety
CEN/TC 278/WG 16	Cooperative ITS
CEN/TC 278/WG 17	Urban ITS
CEN/TC 319	Maintenance
CEN/TC 319/WG 4	Terminology
CEN/TC 319/WG 6	Maintenance performance and indicators
CEN/TC 319/WG 7	Maintenance of buildings
CEN/TC 319/WG 8	Maintenance functions and maintenance management
CEN/TC 319/WG 9	Qualification of personnel
CEN/TC 319/WG 10	Maintenance within physical asset management
CEN/TC 319/WG 11	Condition assessment methodologies
CEN/TC 319/WG 12	Risk-based inspection framework (RBIF)
CEN/TC 319/WG 13	Maintenance process
CEN/TC 348	Facility management
CEN/TC 348/WG 1	Facility management – terms and definitions
CEN/TC 348/WG 4	Taxonomy of facility management
CEN/TC 348/WG 5	Processes in facility management
CEN/TC 348/WG 6	Space measurement in facility management
CEN/TC 348/WG 7	Guide to benchmark for facility management
CEN/TC 411	Bio-based products
CEN/TC 439	Private security services
CEN/TC 439/WG 1	Critical infrastructure protection (CIP)
CEN/TC 447	Horizontal standards for the provision of services
CEN/TC 447/WG 1	Service agreements and contracts
CEN/TC 447/WG 2	Performance management
CEN/TC 447/WG 3	Communications and engagement
CEN/TC 449	Quality of care for older people
CEN/TC 450	Patient involvement in person-centered care

ETUC and ETUI European level trade union organizations are active, and both have offices in Brussels. They directly influence the work of the European Standards Organizations (ESOs). The contact should be with ETUC for services, industrial advancement, or general standardization matters. If the involvement is in the area of machinery and ergonomics, ETUI should be contacted.

Citizens and workers can also participate on global international level via:

- International Organization for Standardization (ISO) [6]
- International Electrotechnical Commission (IEC) [16]

The third important topic, the environmental sustainability, is of high interest for all people on this planet. A considerable interaction exists between environment and standards. Standards affect environment and its quality and sustainability in various ways. They set minimum environmental performance levels of goods and processes. Standards define measurement methodologies to monitor environmental quality of human environments. They can also be developed to affect labeling requirements for products such as domestic appliances. In Europe, European Commission recognizes European Environmental Citizens' Organization for Standardization, ECOS. ECOS provides input to European standards of the three ESOs with the mandate from the EC (CEN [12], CENELEC [13] and ETSI [17]). It advances societal and environmental protection in standards' development on behalf of European environmental nonprofit organizations.

Even though standards are voluntary documents, laws and regulations may make compliance to them compulsory. In this way, the role of standards changes from voluntary to compulsory documents. In the area of environment, energy-using and energy-related products, a good example is Eco-Design Directive for Energy-related Products (2009/125/EC) [18] that sets product-specific regulations for home appliances (e.g., TVs, computers, boilers, dishwashers, and washing machines). Requirements for compliance demonstrations test methods are set in the standards. The compliance is required in comparison to the regulations' minimum energy performance requirements. The importance of ECOS' input secured a legal requirement for a 10% efficiency increase in the minimum energy performance of air conditioners sold in the EU from 2013. ECOS played a key role in CEN mandatory approach development of environmental issues in product and service standards, especially in CEN tailored approach of hazardous chemicals in product standards (e.g., some brominated flame retardants in TVs and similar equipment).

ECOS is also active in the field of waste of electrical and electronic equipment (WEEE [19]). This kind of waste is one of the fastest growing in EU. It is calculated that the WEEE amount rises by 30% from 9 Mtons in 2005 to 12 Mtons by 2020. WEEE is a complex mixture of materials and components. One of the reasons why WEEE is so important is that the production of electrical and electronic equipment uses scarce resources but also that it contains hazardous components that can cause major environmental and health problems. The European Commission has issued two important documents on this issue: Directive on waste electrical and electronic

Table 3.3 List of Technical Committees and Working Groups in CENELEC with ECOS participation

Name	Title
CLC/BT	CENELEC technical board
CLC/BTTF 146-1	Losses of small transformers: methods of measurement, marking and other requirements related to eco-design regulation
CLC/TC 2	Rotating machinery
CLC/TC 2/WG 01	Determining losses of AC low voltage motors fed by voltage source converters
CLC/TC 8X	System aspects of electrical energy supply
CLC/TC 8X/WG 01	Physical characteristics of electrical energy (former BTTF 68-6)
CLC/TC 8X/WG 03	Requirements for connection of generators to distribution networks
CLC/TC 8X/WG 04 (DISBANDED)	WG 38 - Endorsement of IEC 60038 as European Standard
CLC/TC 8X/WG 05 (DORMANT)	Smart grid requirements
CLC/TC 8X/WG 06	System aspects for HVDC grid
CLC/TC 13	Electrical energy measurement and control
CLC/TC 13/WG 01	Electricity meters for active energy of class a, b, and c
CLC/TC 13/WG 02	Data models and protocols for additional functionality of and data exchange in interoperable multi-utility smart metering systems
CLC/TC 14	Power transformers
CLC/TC 22X	Power electronics
CLC/TC 22X/WG 02	Power drive systems
CLC/TC 22X/WG 04	Uninterruptable power supply systems
CLC/TC 22X/WG 05	Safety requirements for power semiconductor converter systems (PSCS)
CLC/TC 22X/WG 06 (DORMANT)	Energy efficiency in power drive systems
CLC/TC 22X/WG 07	Power supplies
CLC/TC 23BX	Switches, boxes, and enclosures for household and similar purposes, plugs and socket-outlets for DC and for the charging of electrical vehicles including their connectors
CLC/TC 23BX/WG 02 (DISBANDED)	Switches and related accessories for use in home and building electronic systems (HBES)
CLC/TC 23BX/WG 03	Boxes and enclosures for electrical accessories for household and similar fixed electrical installations
CLC/TC 23BX/WG 06	Luminaire couplers
CLC/TC 23BX/WG 10	Accessibility
CLC/SR 23	Electrical accessories
CLC/SR 23B	Plugs, socket-outlets and switches
CLC/SR 23H (DISBANDED)	Industrial plugs and socket-outlets
CLC/SR 34B (DISBANDED)	Lamp caps and holders
CLC/TC 34A	Lamps
CLC/TC 34Z	Luminaires and associated equipment
CLC/TC 38	Instrument transformers
CLC/TC 57	Power systems management and associated information exchange
CLC/TC 59X	Performance of household and similar electrical appliances

(continued)

Table 3.3 (continued)

Name	Title
CLC/TC 59X/WG 01	Laundry appliances
CLC/TC 59X/WG 02	Dishwashers
CLC/TC 59X/WG 04	Water heaters
CLC/TC 59X/WG 06	Surface cleaning appliances
CLC/TC 59X/WG 07	Smart household appliances
CLC/TC 59X/WG 08	Performance of electrical household and similar cooling and freezing appliances
CLC/TC 59X/WG 10	Surface cooking appliances
CLC/TC 59X/WG 11	Power consumption of vending machines
CLC/TC 59X/WG 12	Electric room heating appliances
CLC/TC 59X/WG 14	Measurement of noise emission
CLC/TC 59X/WG 15	Coffee makers
CLC/TC 59X/WG 16	Uncertainties and tolerances
CLC/TC 59X/WG 17	Ovens
CLC/TC 59X/WG 20	Commercial refrigerating appliances for use in commercial kitchens
CLC/TC 61	Safety of household and similar electrical appliances
CLC/TC 61/JWG CENTC182TC61	Refrigeration safety (JWG CEN/TC 182 CLC/TC 61)
CLC/TC 69X	Electrical systems for electric road vehicles
CLC/TC 69X/WG 01	AC charging
CLC/TC 69X/WG 02	DC charging
CLC/TC 69X/WG 03	Inductive charging
CLC/TC 69X/WG 04	EMC
CLC/TC 69X/WG 05	Light electric vehicles
CLC/TC 100X	Audio, video and multimedia systems and equipment and related sub-systems
CLC/TC 108X	Safety of electronic equipment within the fields of audio/video, information technology and communication technology
CLC/TC 111X	Environment
CLC/TC 111X/WG 04 (DISBANDED)	End of life requirements for household appliances containing volatile fluorinated substances or volatile hydrocarbons
CLC/TC 111X/WG 05	CLC adoption of IEC TR 62476 as EN standard
CLC/TC 111X/WG 06	WEEE recycling standards
CLC/TC 111X/WG 07	Development of a proposal for EN 50614 "Requirements for the preparing for re-use of waste electrical and electronic equipment"
CLC/SR 118	Smart grid user interface
CLC/TC 205	Home and building electronic systems (HBES)
CLC/SC 205A	Mains communicating systems
CLC/TC 205/WG 18	Smart grids
CEN/CLC/TC 10	Energy-related products – material efficiency aspects for ecodesign
CEN/CLC/TC 10/WG 1	Terminology
CEN/CLC/TC 10/WG 2	Durability
CEN/CLC/TC 10/WG 3	Upgradability, ability to repair, facilitate re-use, use or re-used components
CEN/CLC/TC 10/WG 4	Ability to re-manufacture
CEN/CLC/TC 10/WG 5	Recyclability, recoverability, RRR index, recycling, use of recycled materials
CEN/CLC/TC 10/WG 6	Documentation and/or marking regarding information relating to material efficiency of the product

equipment (i.e., WEEE Directive, 2012/19/EU [19]) and Directive on the restriction of the use of certain hazardous substances in electrical and electronic equipment (i.e., RoHS Directive, 2011/65/EU [20]). These Directives were issued to improve WEEE environmental management by substitution of heavy metals such as lead, mercury, cadmium, and hexavalent chromium and flame retardants such as polybrominated biphenyls (PBB) or polybrominated diphenyl ethers (PBDE) with safer alternatives.

On the national level, for example, in Germany, Koordinierungsbüro Normungsarbeit der Umweltverbände (KNU) [21] takes care of environmental interest representation in standards development.

Table 3.3 shows the list of Technical Committees and Working Groups in CENELEC with ECOS participation. The number of 74 TCs and WGs is definitely very high. It can be assumed only that the number will grow, due to the engagement of the EU in the area of environment.

Bibliography

1. OECD/Eurostat, Oslo manual 2018: guidelines for collecting, reporting and using data on innovation, in *The Measurement of Scientific, Technological and Innovation Activities*, 4th edn., (OECD Publishing, Paris/Eurostat, Luxembourg, 2018)
2. ISO 9000, Internet page: www.iso.org/iso-9001-quality-management.html. Retrieved 19.8.2019
3. OECD, *The Innovation Imperative, Contributing to Productivity, Growth and Well-Being* (OECD Publishing, Paris, 2015)
4. OECD, *The OECD Innovation Strategy: Getting a Head Start on Tomorrow* (OECD Publishing, Paris, 2010)
5. OECD, *The Future of Productivity* (OECD Publishing, Paris/Eurostat, Luxembourg, 2015)
6. ISO, Internet page: www.iso.ch. Retrieved 20.8.2019
7. Regulation (EU) No 1025/2012 of the European Parliament and of the Council of 25 October 2012 on European standardisation, amending Council Directives 89/686/EEC and 93/15/EEC and Directives 94/9/EC, 94/25/EC, 95/16/EC, 97/23/EC, 98/34/EC, 2004/22/EC, 2007/23/EC, 2009/23/EC and 2009/105/EC of the European Parliament and of the Council and repealing Council Decision 87/95/EEC and Decision No 1673/2006/EC of the European Parliament and of the Council, OJ L 316/12
8. European consumer voice in standardization, ANEC, Internet page: www.anec.eu. Retrieved 3.8.2019
9. European Trade Union Confederation, ETUC, Internet page: www.etuc.org/en. Retrieved 3.8.2019
10. European Environmental Citizens Organization for Standardization, ECOS, Internet page: ecostandard.org. Retrieved 3.8.2019
11. European Trade Union Institute, ETUI, Internet page: www.etui.org. Retrieved 4.8.2019
12. CEN National Standards Body, Internet page: www.cen.eu. Retrieved 7.8.2019
13. CENELEC National Committee, Internet page: www.cenelec.eu. Retrieved 8.8.2019
14. EUROGIP, Internet page: www.eurogip.fr/en. Retrieved 11.8.2019
15. KAN, Internet page: www.kan.de/en/what-we-do/. Retrieved 12.8.2019

16. International Electrotechnical Commission (IEC), Internet page: www.iec.ch. Retrieved 2.8.2019
17. ETSI, Internet page: www.etsi.org. Retrieved 31.7.2019
18. Directive 2009/125/EC of the European Parliament and of the Council of 21 October 2009 establishing a framework for the setting of ecodesign requirements for energy-related products (2009/125/EC, OJ L285/10)
19. Directive 2012/19/EU of the European Parliament and of the Council of 4 July 2012 on waste electrical and electronic equipment (WEEE Directive), OJ L 197, 24.7.2012, p. 38–71
20. Directive 2011/65/EU of the European Parliament and of the Council of 8 June 2011 on the restriction of the use of certain hazardous substances in electrical and electronic equipment (RoHS Directive), OJ L 174, 1.7.2011, p. 88–110
21. Koordinierungsbüro Normungsarbeit der Umweltverbände (KNU), Internet page: www.knu.info. Retrieved 3.8.2019

Chapter 4
Innovation and Technical Standardization Documents

4.1 Innovation Documents

The intellectual property is an important driver of the innovation that develops employment. Therefore, it is of utmost value for any country and/or supranational union (e.g., European Union, EU) to take excellent care of the Intellectual Property Rights (IPR) for serving the knowledge-based economy, as well as for avoiding trade barriers. IPR protect intangible assets, thus allowing IPR owners to develop even more their innovative activities.

In the EU, several measures are published with the aim of improvement of IPR enforcement. These are three communications, one report, and one report with a study, as follows:

- Communication on the balanced IP enforcement system in today's society [1]
- Communication providing guidance and clarifying certain provisions of the IPR Enforcement Directive to ensure a more homogeneous interpretation in Europe [2]
- Communication on the EU approach setting to standard essential patents [3]
- Overview report on the functioning of the Memorandum of Understanding on the sale of counterfeit goods via the internet [4]
- An evaluation report and study on the Directive on the enforcement of IPR [5]

Intellectual Property Rights generally refer to the rights associated with intangible knowledge.

IPR can take a different form. It can be:

- A patent
- A utility model
- A copyright

The three forms of IPR can be seen on Fig. 4.1.

D. Šimunić, I. Pavić, *Standards and Innovations in Information Technology and Communications*, https://doi.org/10.1007/978-3-030-44417-4_4

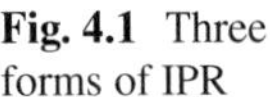

Fig. 4.1 Three forms of IPR

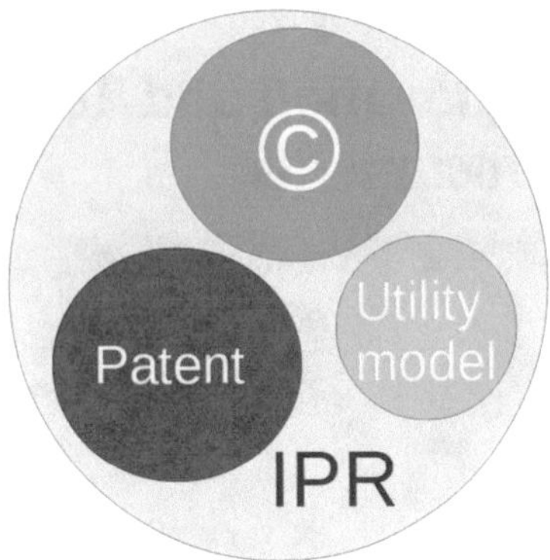

A patent is a grant of a right to the inventor, usually of technical nature. Applications for patents have to be filed with a national or supranational patent office. The applications are examined with the result of giving the grant or refusal to the patent. The patent holder (inventor) has the right to exclude others from using the invention for a select time period, which is usually 20 years.

A utility model can be filed in some countries (in EU examples are Germany and Italy). It offers simpler protection that most often means issuing for a shorter time period. Usually, the period is between 3 and 10 years. At most offices, applications are granted without substantive examination and published within a few months. However, the procedures follow the rules and regulations of national intellectual property offices, meaning that the rights are also limited to the jurisdiction of the issuing authority.

A copyright needs no registration, meaning that it is secured automatically from the moment the work was created. A copyright should use the symbol © along with the year of the first publication and the author's name. The copyright protects all intellectual, creative, artistic, and original expression, and not the ideas themselves. For example, in this group are, e.g., software, database, scientific literature, user manuals, ringtones, start-up tone, and images on the cell phone, related to technical field. The period of protection for copyright is quite long: the whole lifetime of the author plus another 50 years (Article 7 of the Berne convention [6] and Article 12 of the TRIPS Agreement [7]). European Directive 2006/116/EC [8] of 12 December 2006, harmonized term of protection of 70 years after the author's death. The European Commission reports that 33 sectors of the EU economy are copyright-intensive with over seven million jobs (3% of EU employment).

The European Union defined set of 11 directives and 2 regulations as a regulatory framework for copyright in order to enable copyright protected goods (e.g., software, etc.) and services (e.g., services offering access to software) free moving within the internal market. These are:

– Directive on the harmonisation of certain aspects of copyright and related rights in the information society ("*InfoSoc Directive*") [9], published on 22 May 2001

- Directive on rental right and lending right and on certain rights related to copyright in the field of intellectual property ("*Rental and Lending Directive*") [10], published on 12 December 2006
- Directive on the resale right for the benefit of the author of an original work of art ("*Resale Right Directive*") [11], published on 27 September 2001
- Directive on the coordination of certain rules concerning copyright and rights related to copyright applicable to satellite broadcasting and cable retransmission ("*Satellite and Cable Directive*") [12], published on 27 September 1993
- Directive on the legal protection of computer programs ("*Software Directive*") [13], published on 23 April 2009
- Directive on the enforcement of intellectual property right ("*IPRED*") [14], published on 29 April 2004
- Directive on the legal protection of databases ("*Database Directive*") [15], published on 11 March 1996
- Directive 2011/77/EU of the European Parliament and of the Council of 27 September 2011 amending Directive 2006/116/EC on the term of protection of copyright and certain related rights [16], published on 27 September 2011
- Directive on certain permitted uses of orphan works ("*Orphan Works Directive*") [17], published on 25 October 2012
- Directive on collective management of copyright and related rights and multi-territorial licensing of rights in musical works for online use in the internal market ("*CRM Directive*") [18], published on 26 February 2014
- Directive on certain permitted uses of certain works and other subject matter protected by copyright and related rights for the benefit of persons who are blind, visually impaired or otherwise print-disabled (*Directive implementing the Marrakech Treaty in the EU*) [19], published on 13 September 2017
- Regulation on the cross-border exchange between the Union and third countries of accessible format copies of certain works and other subject matter protected by copyright and related rights for the benefit of persons who are blind, visually impaired or otherwise print-disabled (*Regulation implementing the Marrakech Treaty in the EU*) [20], published on 13 September 2017
- Regulation on cross-border portability of online content services in the internal market ("*Portability Regulation*") [21], published on 14 June 2017

EU harmonizes also the legal protection of topographies of semiconductor products with one Directive (*Directive 87/54/EC*, [22]) and two Council Decisions (*Council Decision 94/824/EC* [23] and *Council Decision 96/644/EC* [24]). *E-commerce Directive* [25] and the *Conditional Access Directive* [26] contain relevant provisions to the enforcement of copyright.

The overall goal in the EU harmonisation efforts is to enable copyright protected goods (e.g., books, music, films, software, etc.) and services (e.g., services offering access to these goods) to move freely within the internal market. The complete copyright law is monitored by the European Commission and implemented with the help of the *Court of Justice of the European Union (CJEU)* [27].

A database is an arrangement of independent works, data, or other materials in a systematic way. Database is usually accessible by electronic means, i.e., via information and communication technologies. Independent works can be texts, sounds, images, facts, numbers, and data. It is not required that the contents of the database are copyright protected. In Europe, the European Patent Office (EPO) [28] is an international supranational organization with 38 Member States. The European Union Intellectual Property Office (EUIPO) [29] provides access to creative content protected by copyright (via Orphan Work database), as well as helps the authorities in fighting the IP rights infringements by connecting right holders with custom authorities and police (via Enforcement Database, EDB). EUIPO gives on the public disposal collection of comprehensive information. These information include everything from trademarks, designs, owners, representatives, and bulletins (via eSearch plus) over collection of EUIPO decisions to judgments of the General Court, Court of Justice, and national courts (via eSearch Case Law). EUIPO provides also classification and terms for indications of products in each of the official EU languages (EuroLocarno), verification of certified copies (via certified copies), and a collection of trade marks from all participating official trade mark offices which are participating at national, global, and EU level (via TMview). EUIPO provides centralized access point to registered designs information held by all of the participating national offices (via DesignView) and one-step classification gateway to the harmonized database applied in the EU and databases worldwide (via TMclass). The protection of databases is covered by Directive 96/9/EC [30] in EU. The database can be copyright protected only if it is a result of the author's own creative intellectual effort. However, the material that is contained in the database is not protected. The directive introduces "sui generis" right to protect either obtaining, verifying, or presenting the contents of the database. This right prevents others to extract and/or use all or a substantial part of the contents. However, computer programs are excluded from this protection. Copyright for the database lasts as any other copyright. The "sui generis" right is protected for 15 years.

A trade mark signals the origin of products to consumers, meaning that the registration of the new trade mark must be absolutely distinctive from all the others, already existing. EUIPO provides users with information on the alternative media and formats complying with the European Union Trade Mark Regulation (EUTMR) [31]. Three different types of trade marks can be registered: individual marks, certification marks, and collective marks. Individual marks serve to distinguish the goods of one company from the other. Collective marks serve to distinguish the goods of an association of companies from the competitors. Certification marks were introduced recently at EU level (2017). They indicate that goods comply with the certification requirements of a certifying institution, thus showing a sign of supervised quality. Related to the physical properties, there are many different types of trade mark, e.g., word mark, figurative mark, shape mark, sound mark, movement mark, olfactory mark, taste mark, hologram mark, position mark, and tracer mark. Registration of the trade mark can be done by applying the national procedure in the national trade mark offices or by applying the global procedure in the World Intellectual Property Organization (WIPO) [32]. Registration can be also done with the EUIPO to obtain European Union Trade Mark (EUTM) at the whole EU territory. Trade marks can be renewed indefinitely, by adding 10 years of protection for each renewal. But the trade mark can end if there is no genuine use of it after the granted period of 5 years is over. In the first 5 years, there is no need to use the trade mark.

A design specifies how products or their parts look. An industrial design consists of three-dimensional features, such as lines, colors, shapes, contours, materials, textures, etc. Any industrial or handicraft item can have its design protection. Examples are design of composite products, design of single products, sets of articles, graphic symbols, logos, computer icons, drawings and artwork, web designs, maps, etc. Design needs to be novel and to have its individual character to get the protection. Design can have a registered and unregistered protection. Registered protection is done at national level with the relevant national IP office, or at global level with WIPO [32]. In the EU, registration is through the EUIPO [29]. Unregistered design rights are obtained by disclosing them to the general public and for the use. If registered, designs are protected for 5 years, with the maximum of 25 years from the date of filing.

A geographical indication is defined in Article 22(1) of the TRIPS Agreement [7]. It states that "a good" is identified as originating from the territory of a member country with a given quality and reputation of the same good, being the especial attribute due to its geographical origin. At EU level, two terms are defined: Protected Geographical Indication (PGI) and Protected Designation of Origin (PDO).

A plant variety right gives the holder the exclusive right to exploit new plant variety. In order to get the protection, the submission has to be new, distinct, uniform, and stable. The plant variety right is obtained through registration at the International Union for the Protection of New Varieties of Plants (UPOV) or the Community Plant Variety Office for registration of Community plant variety rights throughout the entire European Union. The period may not be shorter than 20 years. For trees and vines, the period may not be shorter than 25 years from the date of the grant.

A semiconductor topography right has the purpose of preventing the copying of original chip designs and the following commercialization. Therefore, the layout design has to be original and not a commonplace among creators of layout designs and integrated circuits manufacturers at the time of their creation. There are no fixed and formal requirements in the TRIPS agreement [7] related to the layout design protection. Therefore, EU Member States can require that filing has to be effected within a certain time period from the date of the first commercial exploitation and that a fee must be paid, as well as that the competent authority provides the obligatory registration of the layout design with the. Finally, disclosure of the information on its electronic function and/or the commercial exploitation somewhere in the world can be required. The granted rights are exclusive. This means that they can prevent others to reproduce, sell or import all of part of the protected design. This right of exploitation expires after 10 years of the first commercial exploitation anywhere in the world or after 10 years of the registration with the competent authority.

A trade secret is based on the confidential business information, provided by an enterprise with a competitive edge. A good example of a trade secret is Coca-Cola, that remains a secret also after 100 years, whereas the patent's life is not more than 20 years. In European Union, Trade Secrets Directive 2016 enables trade secret as a means for business [33].

Website domains are registered through an accredited registrar with top-level extensions, such as generic (.com, .shop, .hotel) or country specific top-level extensions (.de, .fr). Registration fee has to be paid. If the business is in Europe, extensions can also be .eu (availability can be checked at EURid [34]).

Licensing is the way how to exercise the IPR and bring the licensed object to the broader market. A licensing agreement is an agreement for partnership between a licensor (IPR owner) and licensee (IPR authorized user), usually based on a fee or royalty. It can take a form of a technology license agreement, a technology licensing and franchising agreement, and/or a copyright license agreement. IP infringement is any breach of IPR, meaning that the IPR work has been copied or exploited without the permission from IPR owner. Except copying, IPR infringements are "piracy" and "counterfeiting."

International Berne Convention [6], *Rome Convention* [35], WTO *"TRIPS" Agreement* [7], and two World Intellectual Property Organization (*WIPO*) Internet Treaties (the *WIPO Copyright Treaty* [36] and the *WIPO Performances and Phonograms Treaty* [37]) are all reflected in the EU regulatory framework. EU is also signatory of the *Beijing Treaty* on the Protection of Audiovisual Performances [38] and the *Marrakesh Treaty* [39] to Facilitate Access to Published Works for Visually Impaired or otherwise Print Disabled.

4.2 Technical Standardization Documents

ISO [40] defines standardization as "an activity of establishing, with regard to actual or potential problems, provisions for common and repeated use, aimed at the achievement of the optimum degree of order in a given context." Thus, standardization is an activity that gives a standard as the final product. A standard is a document that gives the best knowledge about a subject, being a product, a process, or a service.

Standards are documents (usually with a purchase cost) that can be reached by open access. Experts, who participate in the standards development, perform their work for free to support regulation. In reality, the experts support the product cycle, shown in Fig. 4.2. What is even more important and contributes to the openness of standards is waiving of any rights for IPR for their contribution. Thus, standards have a purchase cost, but without any charge or license fee. In fact, this is the most important part of the standard, because there is absolutely no any worry that somebody's IPR will be infringed. If there is a patented material that is crucial for standard, the patent holder undertaking is necessary. In this case, the patent holder will be willing to negotiate licenses all around the globe on reasonable and nondiscriminatory terms and conditions. If there is no such undertaking, the standards development stops, and thus, the whole time in the process is wasted. In order to prevent this waste, ETSI IPR policy (part of ETSI Directives [41]) balances the interests of IPR holders and the standardization effort. Therefore, ETSI defined so-called FRAND licensing. FRAND means Fair, Reasonable and Non-discriminatory. By applying FRAND principle in licensing, ETSI ensures a fair and adequate reward to the IPR holders for the use of standard essential patents (SEP).

The technical standard is based on the technical knowledge that usually comes from the so-called pre-normative research. Pre-normative research is the research performed before any kind of standardization activities. The aim of pre-normative research is only to demonstrate feasibility of the technology, technique, process, or service (all under the term "product" in the further text) that has to be standardized later on. The other important task of the pre-normative research is to understand the product's limits for the worldwide application. Only after the pre-normative research, it is possible to start the process of developing and preparing a "pre-standard." The examples of pre-standard are Publicly Available Specification (PAS) and Technical Specification (TS). Their main characteristic is that they are in the most cases prepared in the shorter period than classical standard, being mostly only temporary documents. Only after published pre-standard, so-called "co-normative" research can start. The co-normative research is undertaken together with the standardization process with the goals of understanding reproducibility (same user), repeatability (different users), and uncertainty. Therefore, standards are a very important bridge between innovation and commercialization and between research and industry. This bridge appears, because standards make the research results easily accessible in a consistent format to users. This is why the standards support the both: commercialization and innovation. This is also why four cycles of the product development can be defined: innovation, IPR, standardization, and its commercialization (Fig. 4.2). As a matter of fact, they are all in a closed infinite loop.

As illustrated in Fig. 4.2, standards contribute to the commercialization, which further contributes to the innovation by launching further investments and giving

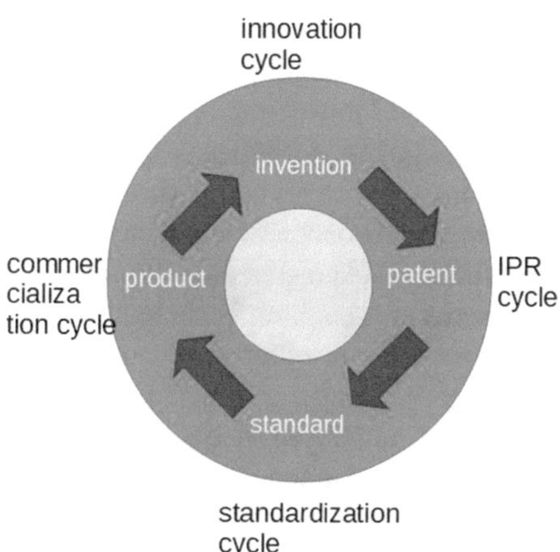

Fig. 4.2 Four product cycles: from the invention to the product; from the innovation process to commercialization process

feedback from customers. Standards also strongly contribute to the validation process of the new measurement tools and methods. Standards are very important in the society shaping, as well as for ensuring safety, quality, and reliability of goods: goods, processes, and services. Finally, the standards ensure the best possible production. Exactly because of the efficient production and the global area of influence, standards perform very important action of enabling serious cost reduction through competition on the market, as well as the interoperability of the goods.

Standards work with the "state of the art," or, as it is stated in the ISO definition [40], with the "optimum degree of order" of the existing and available knowledge. This means that the developed standard is based on the current best available and consolidated knowledge from science, technology, and experience.

A normative document on a technical subject constitutes an acknowledged rule of technology reflecting the state of the art at the time of its approval by a majority of representative experts.

Standardization aims can be various. One of the standardization aims is that a good, process, or service is fit for its purpose, meaning that it is able to serve a defined purpose under specific and given conditions. The other is that a good, process, or service is compatible, meaning that it is suitable for a joint use under specific conditions without causing unacceptable interactions. The third aim is that a good, process, or service is interchangeable, so that it is used in place of another for fulfilling the same requirements. Interchangeability can be functional or dimensional, depending on the aspect. Variety control, as the fourth aim, is a selection of the optimum number of sizes or types of goods, processes, or services for meeting prevailing needs, thus dealing with variety reduction. Another aim is safety, defined as a freedom from unacceptable risk of harm for humans and goods. Standardization's aim is also to achieve environmental protection as a preservation from unacceptable damage from operations of goods, processes, and services during its use, transport, or storage. All these aims are not exclusive, and they can be overlapping.

4.2.1 Normative Documents

A normative document is a generic term for a document or, better to say, any medium with information recorded on or in it, which provides rules, guidelines, or characteristics for activities or their results. The terms for different kinds of normative documents are related to the document's content as a single entity.

A technical document is a document, which gives technical requirements as rules, guidelines, or characteristics for activities or their results. Examples of technical documents are standard, Technical Specification, regulation, and code of practice.

A standard is a voluntary document, aimed at the achievement of the highest possible degree of order in a given context. It is based on the joint results of science, technology, and experience. A standard can give rules, guidelines, or characteristics for activities or on the design, use, or performance of materials, goods, processes, services, systems, or persons. It is used for a general and repeated use for optimum community functioning. A standard is agreed by consensus of subject matter experts and approved by recognized standardization organization. A consensus does not imply unanimity.

A consensus is a general agreement, characterized by the absence of sustained opposition to substantial issues by any important part of the concerned interests and by a process that involves seeking to take into account the views of all parties concerned and to reconcile any conflicting arguments.

The status of standards, being voluntary documents, requires no compulsion. This is the main difference to the regulation document that provides binding legislative rules, adopted by an authority, that are obligatory to comply, as given in definition 3.6 [42]. Of course, at the moment when the regulatory document (either national or supranational, e.g., "New approach" Directive of the European Union) refers to the standards, the compliance with the standards starts to be compulsory. This is the reason for the utmost importance of adopting the ENs by the ESOs members and for withdrawing any conflicting national standards, as given in definition 3.2.1.4 [42].

Another case is when the courts decide on liability determination on the basis of the non-compliance to a standard. Usually, this is when there is no relevant regulation.

Standards contain both the normative and informative elements. Normative elements require compliance and use the word "shall." Contrary to normative, informative elements do not use the word "shall," and they give additional clarification, usually in appendices or notes.

If divided according to geography, a standard can be available to global, supranational, national, or provincial public. An International Standard is a standard adopted by a global standards organization, such as International Organization for Standardization (ISO), International Electrotechnical Commission (IEC) [43], or International Telecommunication Union (ITU) [44]. A supranational standard is adopted by a Supranational Standardization Organization, such as European Committee for Standardization (CEN) [45], European Committee for Electrotechnical Standardization (CENELEC) [46], or European Telecommunications Standards Institute (ETSI) [47] in Europe, Pacific Area Standards Congress (PASC), the Pan American Standards Commission (COPANT), the African Organization for Standardization (ARSO), the Arabic Industrial Development and Mining Organization (AIDMO), and others. A national standard is adopted by a national Standards Body, such as German Institute for Standardization (DIN) in Germany [48]. Provincial standard is adopted at the level of a territorial division of a country. The examples of the territorial division are municipality or city, or in the industry, it can be an association or a company, including individual factories, workshops, or offices.

All the mentioned kinds of "geography" standards are formal standards. The term formal standard relates to the specification that is developed by independent experts, who are nominated by members of standards organizations and working on a voluntary basis and finally accepted and published by National Standards Bodies (NSBs). The term "de iure (or, as it is sometimes written de jure) standard" relates to a standard that is accepted by usual procedure, most often a formal standard. This is because the term “de iure” (pronunced [de: ju:re]) in Latin language means “in law.” “De iure” standard describes practices that are legally recognized, regardless whether the practice exists or not. In formal standards development, public authorities are the most important stakeholders, especially in the approval phase.

The other stakeholders are business and industry associations, as well as professional bodies. They give considerable contribution. A good example would be standards endorsed by an official standards organization (e.g., American National Standards Institute, ANSI [49]), such as ASCII (American Standard Code for Information Interchange), the most common format for text files in computers and on the Internet.

Formal standards are consensus-based documents that are proposed, developed, and approved by the members of the Standards Body (or in certain cases by a Standards Developing Organization (SDO)). Usually, formal standards development follows the processes and procedures laid out in the Part 1 [50] and Part 2 [51] of ISO/IEC Directives.

Except formal standards, another two types are informal standards and private standards.

Informal standards are developed and published by SDOs like IEEE [52], ASTM [53], SAE [54], SEMI [55], VDI [56], etc.

Private standards are developed by a company or trade association. Private or company standard is described by the term “de facto” standard, from Latin “in fact” or “in effect.” It relates a specification (or protocol or technology), that is widespread whether by right or not and applied without any acceptance of any formal standardization organization. “De facto” standard is usually developed by business and industry associations and professional bodies. This means that “de facto standard” is an industrial standard that has achieved a dominant position by public acceptance or market forces. For example, in the beginning of the interconnection era, the company Hewlett-Packard (HP) developed the revolutionary (at that time!) HPIB standard for the communication between different devices, such as printers, measurement devices, etc. with a computer. Later on this standard was converted to an informal standard: IEEE-488 [57]. The next example is USB for high-speed serial interface in computers and for powering or charging low power external devices (like mobile phones, headphones, portable hard drives) usually using micro-USB plug and socket. Another example is Bluetooth, developed by Ericsson, and later on transposed to an informal standard IEEE standard (IEEE 802.15.1).

As already explained, a pre-standard is a document that serves as a base of a standard in the later development phase. It is adopted provisionally to create necessary application public experience.

Technical Specification is a document defining technical requirements and procedures for goods, processes, or services, such as dimension, labelling, packaging, level of quality, or conformity assessment procedures. It can be completely independent of a standard. But it can also be a standard or a part of a standard.

Code of practice is a document by which practices and/or procedures are recommended for design, manufacture, installation, maintenance, or utilization of equipment, structures, or products.

Regulation is a document adopted by an authority that provides binding legislative rules to ensure certain policy objective, like protection of human health, safety, or environment. There is no requirement for having a consensus to publish the regulation. European Commission publishes regulations.

Technical regulation is a regulation containing technical requirements. By nature, it is mandatory. It either incorporates standard, Technical Specification, or code of practice directly (full of parts of them), or it refers to before-mentioned documents. European Union with Directive (EU) 2015/1535 [58] defines all draft technical regulations:

- Technical Specifications
- Other Requirements
- Rules on Services
- Regulations prohibiting the manufacture, importation, marketing or use of a product or prohibiting the provision or use of a service, or establishment as a service provider

Directive [59] defines "good" as any industrially manufactured product and any agricultural product, including fish products.

"Other requirements" cover requirements for protection of consumers or environment of a product by including its life cycle on the market. This means that conditions of use, reuse, or recycling influence significantly product's composition, nature, and/or its marketing.

There is a big difference between standards and technical regulations. Standards are voluntary, and technical regulations are mandatory. Implications for global trade are very different: the product is not allowed on market if the requirements of a technical regulation are not fulfilled; if the product is non-complying to standards, it is allowed on the market. However, market share may be affected, depending on consumers' preference, related to products that comply to standards. It is common to user global standards as a basis for technical regulations.

World Trade Organization Technical Barriers to Trade (TBT) Agreement [60] specifies that product regulations should describe performance, and not design or descriptive characteristics. As an example, a technical regulation on fire-resistant doors should require successful passing of the door on all the necessary tests on fire resistance. The text could be: "the door must be fire resistant with an x number of minutes burn through time." The text should not be oriented to describe the materials to use to make the product, e.g., that "the door must be made of steel, thick 0.3 meters."

4.2.2 *Function of Standards*

Standards are very important in the human life, but especially in the ICT. From the list of standard types, it is easily understandable what are the most important functions of ICT standards. The following four crucial functions are given here:

- Interoperability and Compatibility
- Quality
- Variety Reduction
- Information and Measurement

Interoperability and compatibility standards are probably the most important for nowadays applied computers and telecommunications systems. Standards are important for safety reasons, especially for the medical devices and, in general, for the whole medical world. Thus, it can be said that standards ensure quality that is in the medical world ensuring life. Standards also bring optimization of variety reduction, enabling prices to reach the minimum point. Finally, standardized measurement and testing methods for a detailed description and evaluation of the product's main physical and functional attributes enable homogeneous comparison of the whole market and, thus, easier improvement of the final worldwide product.

4.2.3 Standard Types

Common types of standards are as follows:

- Basic Standard
- Terminology Standard
- Testing Standard
- Product Standard
- Process Standard
- Service Standard
- Interface Standard
- Standard on Data to be Provided

Basic standard is a standard with a wide-ranging coverage or containing general provisions for a particular field. Basic standard can be a basis for other standards, or it can be a standard for direct application. Terminology standard is a standard that deals with terms. It is usually accompanied by the terms' definitions. Sometimes it is accompanied by explanatory notes, illustrations, examples, etc. Testing standard is a standard dealing with test methods. It can be supplemented with other provisions related to testing (sampling, use of statistical methods or sequence of tests). Product standard is a standard defining requirement that a product or a group of products have to fulfill. It may include terminology, sampling, testing, packaging and labelling, and, sometimes, processing requirements. If it specifies only a part of the necessary requirements, it can be dimensional, material, and/or technical delivery standard. Service standard is a standard specifying requirements for a service to be fulfilled, especially in the field of telecommunications, banking, trading, or transport. Interface standard is a standard dealing with requirements for compatibility of products or systems at points of their interconnection. Standard on data to be provided is a standard containing a list of characteristics for which values or other data are to be stated for specifying the good, process, or service. Some data are stated by suppliers, others by purchasers.

As explained above, a particular product standard can be seen also as a testing standard if it provides test methods for characteristics of the product in question.

In today's ICT, possibly the most important goal is to achieve interoperability in multi-network, multiservice, and multi-vendor environment. This is why a lot of attention is given to the standards and the standardization process.

Harmonized or equivalent standards are standards dealing with the same subject but approved by different standardizing bodies. Usually, they establish goods, processes, and services' interchangeability. They deal with mutual understanding of provided test results and/or information. However, harmonized standards might be different even in substance. For example, explanatory notes, guidance on fulfilling the requirements of the standard, and preferences for alternatives and varieties could be different. Unified standards are harmonized standards with the identical substance but different presentation. Identical standards are harmonized standards with the identical substance and presentation. If they are in different languages, identical standards are accurate translations. Internationally harmonized standards are harmonized with a global standard. Bilaterally harmonized standards are harmonized between two standardizing bodies. Multilaterally harmonized standards are harmonized between more than two standardizing bodies. Unilaterally aligned standard is aligned, but not harmonized, with another standard to meet the requirements of the former standard for goods, processes, services, tests, and information. Since it is unilateral, vice versa alignment does not exist. Comparable standards are standards approved by different standardizing bodies on the same goods, processes, or services, permitting unambiguous comparison of differences in the requirements. This is possible, because they contain different requirements on the same characteristics and assessed by the same methods. Comparable standards are not harmonized (or equivalent) standards.

Technical regulations can be harmonized like standards. It is necessary to replace "standards" by "technical regulations" and "standardizing bodies" by "authorities."

4.2.4 Contents of Normative Documents

Provision is an expression in the content of a normative document that takes the form of a statement, an instruction, a recommendation, or a requirement.

A provision that conveys information is a statement. Instruction is a provision that conveys an action to be performed. Recommendation conveys advice or guidance and requirement criteria to be fulfilled. Exclusive requirement is a mandatory requirement (deprecated) of a normative document that has to be fulfilled for a compliance with that document. "Mandatory requirement" is the only one that is compulsory by a law or a regulation. On the other hand, optional requirement of a normative document has to be fulfilled for a compliance with a particular option permitted by that document. For example, it can be an additional requirement that must be fulfilled only if applicable and that may otherwise be disregarded. Deemed-to-satisfy provision indicates one or more means of compliance with a requirement of a normative document. Descriptive provision concerns the characteristics of a good, process, or service, and it usually describes design, constructional details, etc. with dimensions and material composition. Performance provision concerns behavior of a good, process, or service, or it is related to their use. The above provision types use different wording form: instructions are expressed in the imperative mood, recommendations by the use of the auxiliary "should," and requirements by the use of the auxiliary "shall."

4.2.5 Structure of Normative Documents

A normative document usually consists of a body and additional elements. The body is a set of provisions that comprises the substance of a normative document. If that normative document is a standard, the body comprises general elements related to its subject and definitions and main elements conveying provisions. Additional elements can be normative annexes or informative annexes. Normative annexes contain parts of the normative document body, whereas informative annexes do not have any of the contents of the normative document body and do not have any effect on its substance. If the normative document is a standard, additional elements may be, e.g., details of publication, foreword, or notes.

4.2.6 Implementation of Normative Documents

A normative document can be applied in two different ways. One way is in a production or in a trade. It can be also taken over, in part or as another (the second) normative document. The second document can be directly applied or taken again over to another normative document. Taking over a global standard (in a national normative document) means that a national normative document is based on the relevant global standard. It can also mean that a global standard is endorsed into a national normative document, with some identified deviations from the global standard.

Application of a normative document means use of a normative document in production, trade, etc. It can be direct, meaning that international/supranational/global standard is applied irrespectively of taking over to any other normative document. It can be also indirect, meaning that International Standard is applied through the medium of another normative document in which it has been taken over.

4.2.7 *References to Standards in Regulations*

A reference to one or more standards is done in place of detailed provisions within a regulation. It can be dated, undated or general. At the same time, it can be exclusive or indicative. It can be also linked to a more general legal provision referring to the state of the art or acknowledged rules of technology, as a stand-alone provision. Dated reference to standards identifies one or more specific standards so that later standard revisions do not apply unless the regulation is modified. The standard is usually identified primarily by its number. Important are also date or edition, title, and the organization that issued it. Undated reference to standards identifies one or more specific standards so that later standard revisions can be applied without the need to modify the regulation. General reference to standards designates all standards of a specified body and/or in a particular field without individual identification. Exclusive reference in a regulation or a general law can be made standard compulsory, and then it becomes mandatory standard.

4.2.8 *Global (International) Normative Documents*

The three most important global standardization organizations are ISO, IEC, and ITU.

(a) ISO

ISO [40] publishes various types of deliverables. These are:

- ISO Standards
- ISO/TS Technical Specifications
- ISO/TR Technical Reports
- ISO/PAS Publicly Available Specifications
- IWA International Workshop Agreements
- ISO Guides

ISO Standards or better known as ISO International Standards are written with the goal to achieve the best possible degree of order in a certain context by providing rules or guidelines activities. Various types of standards exist, for example, product standards, test methods, codes of practice, and guideline standards. ISO is famous for management systems standards series. As of May 2019, ISO published 22,631 standards.

ISO/TS Technical Specification is a document that is published in a very short time period in comparison with International Standard. It is meant to support work under technical development with a strong future. It is published with the intention to be applicable for immediate use, but that it will be eventually transformed to ISO Standard.

ISO/TR Technical Report is an informative document, meaning that it gives, for example, state of the art of certain technology or of a sector.

ISO/PAS Publicly Available Specification is a document that is published for immediate use and for possible transformation to an ISO Standard. A group of experts of a technical working group or of an organization external to ISO can initiate drafting and publishing PAS, because of an urgent market need.

A member body typically supports International Workshop Agreement (IWA). It is a document developed outside the classical ISO committee system for enabling "open workshop" environment. It has maximum lifespan of six years. If it is not meanwhile transformed to some other ISO deliverable, it is automatically withdrawn.

ISO Guide is a document, written for an enhancement of understanding about the certain area.

(b) IEC

IEC [58] publishes various types of deliverables. These are:

- IEC International Standards, IS
- IEC/TS Technical Specifications
- IEC/TR Technical Reports
- IEC/PAS Publicly Available Specifications
- IEC Amendments
- IEC Technical Corrigenda
- IEC Interpretation Sheets
- IEC Guides

IEC publishes International Standards (ISs) in the field of electrotechnology, for all devices using, producing, or storing electricity or containing any kind of electronics. IEC fields of activity are quite wide: from nanotechnology through electronic components, including all kinds of batteries and fuel cells as well as household devices. Finally, they cover also all kinds of vehicles and electric power generation and transfer.

IEC International Standard is a result of the consensus process of all the national members representatives (National Committee, NC), meaning that at least 75% of the participants agreed with the standard contents. The initiative for any standard comes from the NCs, representing country interests with or without industry bias. Each country has only one vote, independent of the country size. If interested, companies participate in the standards development through their NCs. The participation in the development, as well as the standard use, is voluntary. Any liaison international organization, broad supranational organization, consortium, and forum may participate in the standard preparation.

IEC publishes Technical Specification (TS) in the case of premature standardization of certain subject or in the case of insufficient consensus required for the IS. TS is as detailed and complete document as it is the IS. Sixty-seven percent of

Participating Members (P-Member) of an IEC Technical Committee (TC) or Subcommittee (SC) is required for approval of TS.

IEC Technical Report (TR) is a fully informative document that does not contain any normative matter, e.g., that could be data on "state of the art" in relation to standards of national committees on a particular subject.

IEC Publicly Available Specification (PAS) has an aim to speed up standardization due to an urgent market need. Competing PASs in the same area are permitted, since PAS can result out of a consensus of experts within the same IEC Working Group (WG) or out of a consensus of manufacturers or commercial associations, industrial consortia, user group, and professional and scientific societies, external to the IEC. The committee can designate PAS as a Pre-standard.

IEC Amendment is a normative document that is approved by the consensus of the IEC Membership. IEC Amendment changes the technical normative element(s) of an International Standard.

IEC Technical Corrigendum is issued to correct a technical error or ambiguity in an International Standard, a Technical Specification, a Publicly Available Specification, or a Technical Report. Technical Corrigendum is issued only if the error or ambiguity could lead to an incorrect or unsafe application.

IEC Interpretation Sheet is a formal explanation to a direct or indirect urgent request of a standards user, such as certification body, manufacturer, or testing laboratory. Indirect request goes usually via an IEC conformity assessment scheme.

IEC Guide defines rules, orientation, advice, or recommendations related to global standardization and conformity assessment.

(c) ITU

International Telecommunication Union [60] has three sectors: ITU-R, ITU-T, and ITU-D. ITU-R stands for ITU Radiocommunication Sector, ITU-T for ITU Telecommunication Standardization Sector, and ITU-D for ITU Telecommunication Development Sector.

ITU-R publishes ITU-R Recommendations, ITU-R Questions, ITU-R Reports, ITU-R Resolutions, and ITU-R Handbooks.

ITU-T generates ITU-T Recommendations but also documents that serve the membership such as the following:

Telecommunication Standardization Bureau (TSB) Circulars are letters that are circulated to all participants and contain information of general interest such as the meeting schedule or work procedures. TSB Circulars are open public documents.

TSB Collective letters are sent from SGs to the registered participants of certain SG. They are also used to collect information to SG. TSB Collective letters are open public documents.

ITU-T also posts some meeting documents on SGs' web pages, such as contributions, reports, ITU Operational bulletin, Handbooks and guides, temporary documents, and liaison statements.

Contributions are submitted by Sector participants to TSB, and they can be dispatched to the Sector participants registered in the relevant SG. Contributions can be accessed via TIES account.

Reports are dispatched after a meeting to Sector participants registered in the relevant SG. They are accessible via TIES account.

Temporary Documents contain reports from Chairmen, Rapporteurs, or Drafting Groups. They are distributed only to the present participants. They are accessible via TIES account.

Liaison Statements transmit information from one group to another. They are distributed only to the participants present and are accessible via TIES account.

ITU-D publishes reports (e.g., about ICT in the least developed countries), yearbook of statistics for telecommunications/ICT, guides (e.g., Guide to developing a national cybersecurity strategy), etc.

4.2.9 Supranational Normative Documents

Normative documents at supranational level will not be studied in general, but rather on the example of the European Union, as an example of the successful transnational unity.

(a) CEN

CEN [55] delivers the following:

- European Standards (EN)
- Technical Specifications (TS)
- Technical Reports (TR)
- Guides (CG)
- CEN and/or CENELEC Workshop Agreements (CWA)
- Pre-standards (ENV)
- Reports (CR)
- European Standards identical to International Standards (ISO)

A variety of deliverables exist due to the various market needs and, thus, development methods, approval processes, and implementation, including development time.

The highest effort of CEN goes to preparation and development of European Standards (EN). Technical Committees (TCs) prepare standards. Each TC has its own scope or field of operation within which a work program of identified standards is developed. TC works on the basis of national participation by the CEN members, reflecting an achieved wide consensus of European Standards. TC creates subcommittee in the case of a large work program. Working Groups (WGs) mostly develop the technical standard. Experts, appointed by the CEN Members, develop a draft standard. When adopted, European Standard has automatically the status of the national standard in all CEN Member Countries, meaning that all the conflicting national standards in the mentioned CEN countries have to be withdrawn before the final adoption. All the ENs are published in *Official Journal of the European Communities* [61].

CEN Technical Specification (CEN/TS) can be used in CEN TC as European Pre-standard (ENV) for innovative characteristics of the new technology. Anticipation of the future harmonization in the situation with co-existence of several technological alternatives can also be brought as CEN/TS. CEN/TS does not go through the same procedure as national standards, and thus it does not have the status of European Standard.

CEN Workshop Agreement (CWA) is a standardization document, which is open to the direct participation of all interested parties. Documents CWA are published quickly (typically between 10 and 12 months), which means that they can significantly contribute to the fast delivery of the market needs. CWA is a result of the CEN Workshop that is open to all interested parties. The participation is not limited to the CEN Members, which contributes to the value of the document. The CWA can serve as an alternative way toward formal European Standard. Also, CWA can serve as a direct way to come to the ISO level. This is a non-formal document, so there is no need to adopt it in the CEN Member Countries.

Workshops are organized when there is a need to quickly react to rapidly changing technology in a form of specification. Their result is CEN Workshop Agreements (CWA).

CEN is one of the three official standardization bodies of EC (so-called European Standards Organization (ESO)). It prepares standards in support to the "New approach," according to the given mandate of the European Commission.

CEN produces a lot of documents. For example, in 2018, there were 1198 of them: 1076 European Standards, 58 Technical Specifications, 44 Technical Reports, 18 CEN Workshop Agreements, and 2 CEN Guides (CGs).

(b) CENELEC

CENELEC publishes:

- European Standard, EN
- Harmonization Document, HD

These two documents are called "standards" and have to be applied by all CENELEC country members, which means that all the other conflicting standards have to be withdrawn. There is an important difference between EN and HD. EN is taken as it is, and the technical contents from HD should be taken, irrespective of an exact wording or of a number of documents that are containing those contents.

EN is a standard that is available in three languages of CENELEC: English, French, and German. EN is the most important CENELEC product, led by principles of consensus, openness, and transparency. All CENELEC members have obliged themselves for EN use. Before the final issue of EN, all the national members have to give their opinion of the work of certain Working Group or Technical Committee. Only after completion of this, the final document, prepared by Technical Board, can be issued, after which it may be applied by all country members.

The same principles as for EN are valid for HD. There is no obligation to publish the completely exact national standard as HD. It is only important that the technical content of HD is transposed in the equal way in all the documents. CENELEC develops harmonized standards due to the set requests by the European Commission under the "New Approach" European directives area. "New Approach" provides solutions for a compliance with a legal provision in the wide range of electrical engineering area. However, their use is still voluntary. For example, manufacturers or conformity assessment bodies can demonstrate compliance with the mandatory legal requirements also in the other way.

Except EN and HD, CENELEC approves or approved other documents with different goals that follow with their short description. These are:

- Technical Specification, TS
- Technical Report, TR
- Guide, G
- CENELEC Workshop Agreement, CWA

Technical Specification, TS, was a standard document prepared and approved by Technical Committee, and not by CENELEC. There was no need for public opinion. TS had to be written in any of the official languages with the maximum duration of 2 to 3 years. TS was important for developing technologies that would have not be able to collect enough interest from member countries to publish EN. TS should not conflict with any of published CENELEC standard.

Technical Report, TR, was an informative document with technical contents and published in one of the three official languages. TR was approved by Technical Board by majority. There was no time duration, but the recommended version had to be revised from the responsible technical body.

Guide, G, was an informative document that related to the "inner system." It was especially related to the standardization principles. Guide had to be approved on General Assembly or on a Technical Committee level. There is no duration time limit.

CENELEC Workshop Agreement, CWA, is an agreement that is developed according to the agreed and approved consensus during the workshop. Usually, some products of a short-time duration need the open area for pre-standards development in order to shorten the development time. It is published on at least one of the three official languages. Revision is possible.

In addition to CEN, CENELEC is also the official standardization body of EC. It prepares standards in support to the "New Approach," according to the given mandate of European Commission. For example, CENELEC is responsible for preparation of standards in support of the Ecodesign Directive (2009/125/EC) [35] or of the Waste from electrical and electronic equipment directive (WEEE) [36]. All the ENs are published in the *Official Journal of the European Communities* [61].

(c) ETSI

ETSI publishes the following:

- European Standard (EN)
- ETSI Standard (ES)
- ETSI Technical Specification (ETSI TS)
- ETSI Technical Report (ETSI TR)
- ETSI Guide (ETSI EG)
- ETSI Special Report (ETSI SR)
- ETSI Group Specification (ETSI GS)
- ETSI Group Report (ETSI GR)

European Standard (EN) is a document for which European Commission (EC) or European Free Trade Association (EFTA) mandated one of European Standards Organizations (ESOs). EN is written also when specific needs exist in Europe that have to be transposed to national standardization. A technical body that drafts EN is the Technical Committee. ETSI's European National Standards Organization approves it.

Harmonized Standards are ENs that ETSI produces as a response to an EC mandate. EC provides technical details for provision of so-called Essential Requirements of an EC Directive.

Community Specifications are ENs that are also produced in response to an EC mandate in a special area of civil aviation equipment (with European Organization for Civil Aviation Equipment, EUROCAE). As harmonized standards, they are also published in the Official Journal of the European Union [61].

ETSI Standard (ES) contains technical requirements and it requires the full ETSI membership approval.

ETSI Technical Specification (TS) is a technical document, drafted by TC that requires quick availability by users.

ETSI Technical Report (TR) is explanatory document, drafted, and approved by TC.

ETSI Guide (EG) is a guiding document that handles specific technical standardization activities, approved by full ETSI membership.

ETSI Special Report (SR) is an informative document, publicly available, produced and approved by the same TC.

ETSI Group Specification (GS) is a technical and/or explanatory document, produced and approved by Industry Specification Groups (ISGs).

ETSI Group Report (GR) is an informative document, approved by an ISG.

Bibliography

1. Communication, A balanced IP enforcement system responding to today's societal challenges, Internet page: eur-lex.europa.eu/legal-content/EN/TXT/PDF/?uri=CELEX:52017DC0707&from=EN. Retrieved 14.8.2019

2. Communication providing guidance and clarifying certain provisions of the IPR enforcement Directive to ensure a more homogenous interpretation in Europe, Internet page: ec.europa.eu/docsroom/documents/26582/attachments/1/translations/en/renditions/native. Retrieved 4.8.2019
3. Communication, Setting out the EU approach to Standard Essential Patents, Internet page: ec.europa.eu/docsroom/documents/26583. Retrieved 13.8.2019
4. Report from the Commission to the European Parliament and the Council on the functioning of the Memorandum of Understanding on the Sale of Counterfeit Goods via the Internet, COM/2013/0209 final, Internet page: eur-lex.europa.eu/legal-content/EN/TXT/PDF/?uri=CELEX:52013DC0209&from=EN. Retrieved 11.8.2019
5. Support study for the ex-post evaluation and ex-ante impact analysis of the IPR enforcement Directive (IPRED), Directorate-General for Internal Market, Industry, Entrepreneurship and SMEs (European Commission), EY, Schalast, Technopolis, 2017, Internet page: publications.europa.eu/en/publication-detail/-/publication/29df5fb0-d0cc-11e7-a7df-01aa75ed71a1/language-en/format-PDF/source-search. Retrieved 16.8.2019
6. Berne convention, Internet page: www.wipo.int/treaties/en/ip/berne/. Retrieved 15.8.2019
7. TRIPS Agreement, The Agreement on Trade-Related Aspects of Intellectual Property Rights (TRIPS), Internet page: wto.org/english/docs_e/legal_e/27-trips.pdf. Retrieved 20.8.2019
8. Directive 2006/116/EC of the European Parliament and of the Council of 12 December 2006 on the term of protection of copyright and certain related rights (codified version), Internet page: eur-lex.europa.eu/legal-content/EN/ALL/?uri=CELEX:32006L0116. Retrieved 9.8.2019
9. Directive on the harmonisation of certain aspects of copyright and related rights in the information society ("InfoSoc Directive"), 22 May 2001, Internet page: eur-lex.europa.eu/legal-content/EN/TXT/?uri=CELEX:32001L0029. Retrieved 11.8.2019
10. Directive on rental right and lending right and on certain rights related to copyright in the field of intellectual property ("Rental and Lending Directive"), 12 December 2006, Internet page: eur-lex.europa.eu/legal-content/EN/ALL/?uri=CELEX:32006L0115. Retrieved 10.8.2019
11. Directive on the resale right for the benefit of the author of an original work of art ("Resale Right Directive"), 27 September 2001, Internet page: eur-lex.europa.eu/legal-content/EN/TXT/?uri=celex:32001L0084. Retrieved 10.8.2019
12. Directive on the coordination of certain rules concerning copyright and rights related to copyright applicable to satellite broadcasting and cable retransmission ("Satellite and Cable Directive"), 27 September 1993, Internet page: eur-lex.europa.eu/legal-content/EN/ALL/?uri=CELEX:31993L0083. Retrieved 10.8.2019
13. Directive on the legal protection of computer programs ("*Software Directive*"), 23 April 2009, Internet page: eur-lex.europa.eu/legal-content/en/TXT/?uri=CELEX:32009L0024. Retrieved 5.7.2019
14. Directive on the enforcement of intellectual property right ("*IPRED*"), 29 April 2004, Internet page: eur-lex.europa.eu/legal-content/EN/TXT/?uri=CELEX:32004L0048. Retrieved 9.8.2019
15. Directive on the legal protection of databases ("*Database Directive*"), 11 March 1996, Internet page: eur-lex.europa.eu/legal-content/EN/ALL/?uri=CELEX:31996L0009. Retrieved 6.7.2019
16. Directive 2011/77/EU of the European Parliament and of the Council of 27 September 2011 amending Directive 2006/116/EC on the term of protection of copyright and certain related rights, Internet page: eur-lex.europa.eu/legal-content/EN/TXT/?uri=CELEX:32011L0077. Retrieved 10.8.2019
17. Directive on certain permitted uses of orphan works ("*Orphan Works Directive*"), 25 October 2012, Internet page: eur-lex.europa.eu/legal-content/EN/TXT/?uri=celex:32012L0028. Retrieved 8.7.2019
18. Directive on collective management of copyright and related rights and multi-territorial licensing of rights in musical works for online use in the internal market ("*CRM Directive*"), 26 February 2014, Internet page: eur-lex.europa.eu/legal-content/EN/TXT/?uri=uriserv:OJ.L_.2014.084.01.0072.01.ENG. Retrieved 21.8.2019

19. Directive on certain permitted uses of certain works and other subject matter protected by copyright and related rights for the benefit of persons who are blind, visually impaired or otherwise print-disabled (*Directive implementing the Marrakesh Treaty in the EU*), 13 September 2017, Internet page: eur-lex.europa.eu/eli/dir/2017/1564/oj. Retrieved 14.7.2019
20. Regulation on the cross-border exchange between the Union and third countries of accessible format copies of certain works and other subject matter protected by copyright and related rights for the benefit of persons who are blind, visually impaired or otherwise print-disabled (*Regulation implementing the Marrakech Treaty in the EU*), 13 September 2017, Internet page: eur-lex.europa.eu/legal-content/EN/ALL/?uri=uriserv:OJ.L_.2017.242.01.0001.01.ENG. Retrieved 14.7.2019
21. Regulation on cross-border portability of online content services in the internal market ("*Portability Regulation*"), Internet page: eur-lex.europa.eu/eli/reg/2017/1128/oj. Retrieved 21.8.2019
22. Directive 87/54/EC, Internet page: eur-lex.europa.eu/legal-content/EN/TXT/?uri=CELEX:31987L0054. Retrieved 10.8.2019
23. Council Decision 94/824/EC, Internet page: eur-lex.europa.eu/legal-content/en/ALL/?uri=CELEX:31994D0824. Retrieved 21.8.2019
24. Council Decision 96/644/EC, Internet page: eur-lex.europa.eu/legal-content/EN/TXT/?uri=CELEX:31996D0644. Retrieved 10.7.2019
25. E-commerce Directive, Internet page: eur-lex.europa.eu/legal-content/EN/TXT/?qid=1440754250153&uri=CELEX:32000L0031. Retrieved 4.7.2019
26. Conditional Access Directive, Internet page: eur-lex.europa.eu/legal-content/EN/ALL/?uri=CELEX:31998L0084. Retrieved 12.7.2019
27. Court of Justice of the European Union (CJEU), Internet page: europa.eu/about-eu/institutions-bodies/court-justice/index_en.htm. Retrieved 11.7.2019
28. European Patent Office (EPO), Internet page: www.epo.org. Retrieved 11.7.2019
29. European Union Intellectual Property Office (EUIPO), Internet page: euipo.europa.eu/ohimportal/en. Retrieved 10.7.2019
30. Directive 96/9/EC of the European Parliament and of the Council of 11 March 1996 on the legal protection of databases, Internet page: eur-lex.europa.eu/legal-content/EN/TXT/?uri=celex%3A31996L0009. Retrieved 9.7.2019
31. Regulation (EU) 2017/1001 of the European Parliament and of the Council of 14 June 2017 on the European Union trade mark, *OJ L 154, 16.6.2017, p. 1–99*
32. World Intellectual Property Organization (WIPO), Internet page: www.wipo.int. Retrieved 3.7.2019
33. Directive (EU) 2016/943 of the European Parliament and of the Council of 8 June 2016 on the protection of undisclosed know-how and business information (trade secrets) against their unlawful acquisition, use and disclosure, Internet page: eur-lex.europa.eu/legal-content/EN/TXT/?uri=CELEX%3A32016L0943. Retrieved 14.8.2019
34. EURid, Internet page: www.eurid.eu/en. Retrieved 20.8.2019
35. Rome Convention, Internet page: www.wipo.int/treaties/en/ip/rome/. Retrieved 15.8.2019
36. WIPO Copyright Treaty, Internet page: www.wipo.int/treaties/en/ip/wct/. Retrieved 12.7.2019
37. WIPO Performances and Phonograms Treaty, Internet page: www.wipo.int/treaties/en/ip/wppt/. Retrieved 11.7.2019
38. Beijing Treaty, Internet page: www.wipo.int/treaties/en/ip/beijing/. Retrieved 20.8.2019
39. Marrakesh Treaty, Internet page: www.wipo.int/treaties/en/ip/marrakesh/. Retrieved 21.8.2019
40. ISO, Internet page: www.iso.ch. Retrieved 28.8.2019
41. ETSI Directives, Internet page: portal.etsi.org/docbox/Board/ETSI_Directives/. Retrieved 20.8.2019
42. ISO/IEC Guide 2:2004, Internet page: www.iso.org/standard/39976.html. Retrieved 13.8.2019
43. IEC, Internet page: www.iec.ch. Retrieved 10.8.2019
44. ITU, Internet page: www.itu.int. Retrieved 7.8.2019
45. CEN, Internet page: www.cen.eu. Retrieved 8.8.2019

46. CENELEC, Internet page: www.cenelec.eu. Retrieved 6.8.2019
47. ETSI, Internet page: www.etsi.org. Retrieved 9.8.2019
48. DIN, Internet page: www.din.de. Retrieved 10.8.2019
49. ANSI, Internet page: www.ansi.org. Retrieved 11.8.2019
50. ISO/IEC Directives Part 1 (Procedures for the technical work), Internet page: www.iso.org/iso/standards_development/processes_and_procedures/iso_iec_directives_and_iso_supplement.htm. Retrieved 26.8.2019
51. ISO/IEC Directives Part 2 (Rules for the structure and drafting of international standards), Internet page: www.iso.org/iso/standards_development/processes_and_procedures/iso_iec_directives_and_iso_supplement.htm. Retrieved 26.8.2019
52. IEEE, Internet page: www.ieee.org. Retrieved 10.8.2019
53. ASTM, Internet page: www.astm.org. Retrieved 11.8.2019
54. SAE International, Internet page: www.sae.org. Retrieved 10.8.2019
55. SEMI Standards, Internet page: www.semi.org. Retrieved 11.8.2019
56. VDI Standards, Internet page: m.vdi.eu. Retrieved 11.8.2019
57. IEEE 488.2-1992 – IEEE Standard Codes, Formats, Protocols, and Common Commands for Use With IEEE Std 488.1-1987, IEEE Standard Digital Interface for Programmable Instrumentation, Internet page: standards.ieee.org/standard/488_2-1992.html. Retrieved 10.8.2019
58. Directive (EU) 2015/1535 of the European Parliament and of the Council of 9 September 2015 laying down a procedure for the provision of information in the field of technical regulations and of rules on Information Society services, Internet page: eur-lex.europa.eu/legal-content/EN/TXT/?uri=CELEX%3A32015L1535. Retrieved 11.8.2019
59. Directive 98/71/EC of the European Parliament and of the Council of 13 October 1998 on the legal protection of designs, OJ L 289, 28.10.1998, p. 28–35
60. Technical Barriers to Trade (TBT) Agreement, Internet page: www.wto.org/english/tratop_e/tbt_e/tbt_e.htm. Retrieved 9.8.2019
61. Official Journal of the European Communities, OJEC, Internet page: www.ojec.com. Retrieved 14.6.2019

Chapter 5
Innovation and Technical Standards Life Cycles

5.1 Innovation Life Cycle

The innovation cycle starts with an idea about a goods, service, or process improvement or about a completely new goods, service, or process. The full innovation life cycle is shown in Fig. 5.1. This period has the characteristics of the exploration. Therefore, it is called exploration or identification part. Exploration is a critical starting point, e.g., when stakeholders consider changing their life in a certain way. During exploration, an implementation team assess the potential match between the needs of possible stakeholders, the innovation requirements and stakeholders' resources. In the exploration phase, ideas are collected, derived, evaluated, and finally released into the second part. The design part deals with design of the invention, meaning that various concepts are extensively analyzed. It delivers the concept for the solution, implementation, and marketing. In the third part, which is in many cases related to the standardization, the innovation concept undergoes further development and testing, ending up as the final prototype. The concept may be significantly improved on the international level (easier launch and completion), when it is subjected to the standardization process. The fourth and the final part is the implementation. The finished product is brought to the market, where it is finally "polished" to the customer's needs. Innovation cycles for physical goods, services, and processes are illustrated in Fig. 5.1.

The innovation action rates for all three, goods, services, and processes, are similar. The difference is that there is a delay in time. Whereas the peak of the innovation action rate in the product innovation takes place in the part 2 (design) of the innovation cycle, the peak of the innovation action rate in the service innovation takes place somewhat later, close to the part 3 (standardization). The peak of the innovation action rate in the process innovation takes place even closer to the part 4 (implementation). As time elapses, the innovation action rate drops continuously toward the implementation phase.

D. Šimunić, I. Pavić, *Standards and Innovations in Information Technology and Communications*, https://doi.org/10.1007/978-3-030-44417-4_5

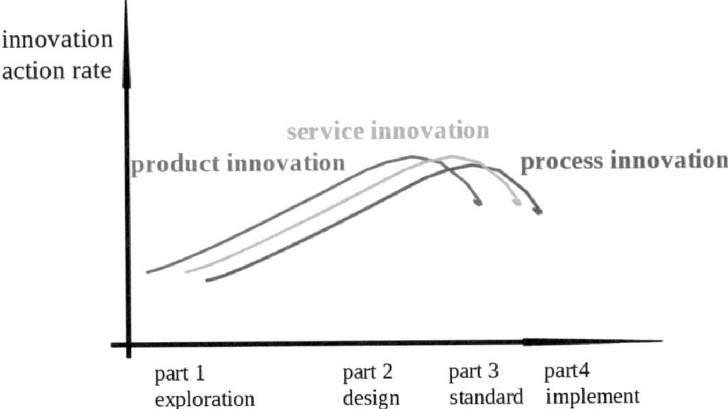

Fig. 5.1 Innovation cycles for goods, services, and processes. All the innovation cycles consist of four parts: (1) exploration; (2) design; (3) standardization; (4) implementation

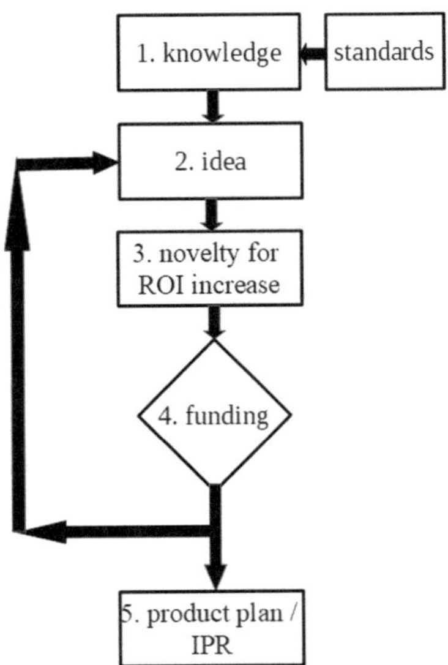

Fig. 5.2 The first part of the innovation process (exploration)

The three parts that are interesting for the innovation are explained in terms of the flowchart, with defined phases in the further text and in Figs. 5.2 and 5.3.

The first part of the innovation process (exploration) consists of answering to the question "What is being innovated?". This implies assembling the knowledge (shown in Fig. 5.2 as phase 1).

An affordable knowledge corpus can be collected from the according standards that cover the desired area of knowledge (phase 1). Only after acquiring the required knowledge, ideas for one or more innovations can start to arise (shown in Fig. 5.2 as phase 2). After establishing the actual novelty, it is necessary to estimate return on investment (ROI) of the novelty (phase 3). Return on investment is defined here as

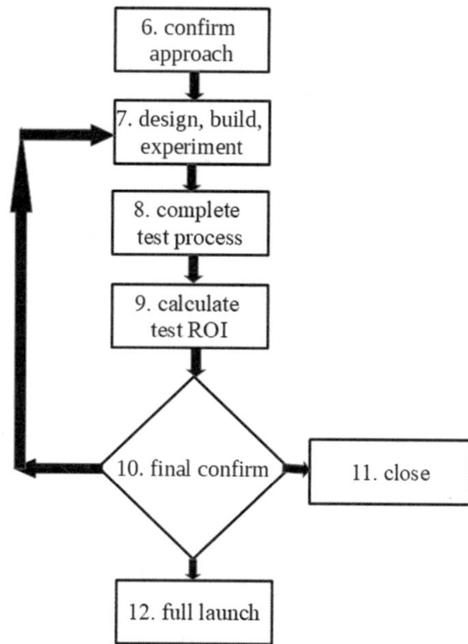

Fig. 5.3 The second part of the innovation process (design)

the ratio of net profit over a period of time and the investment cost. The goal is that the calculated ROI is high. The fourth phase consists of finding the adequate funding (phase 4). If there is no interested investor, the inventor has to think about another idea that will come to the fertile soil related to possible investors. If the funding is found, the next phase (phase 5) consists of a development of the product plan and of a definition whether the inventor(s) and investor(s) will decide to protect Intellectual Property Right (IPR) or not. If yes, the parties should develop the plan for sharing IPR. If there will be no IPR protection, inventor(s) and investor(s) proceed to the second part of the innovation process (design) that answers to the question "How will the idea be implemented?" (Fig. 5.3).

Phase 6 in the second part of the innovation process (design) consists of confirming the taken approach. Phase 7 is about designing, building, and experimenting with the idea that is meanwhile becoming a product. In phase 8, the innovation process is completed, resulting in a test product. In phase 9, the ROI of the test product has to be re-estimated with the finished innovation process. Based on the ROI result, the interest has to be reconfirmed again (phase 10). If reconfirmed, then the final product should be examined for possible improvements with regard to the ROI estimations before completion and full launch (phase 12). However, if the ROI is unsatisfactory, the innovation process stops (phase 11).

After the first part (exploration) and the second part (design), the third part of the innovation cycle is related to the standardization part. In this part, knowledge accumulates and interrelates with other global actors on the innovation scene. The explanation of the standards life cycle follows.

5.2 Technical Standards Life Cycle

As explained in Chap. 4, a technical standards document is established by consensus and approved by a recognized body that provides, for common and repeated use, rules, guidelines, or characteristics for activities or their results, aimed at the achievement of the optimum degree of order in a given context. The approach to approval of the technical standard is different, according to the internal organization of the standardization body. For example, formal standards that are developed in the standardization organization like International Organization for Standardization (ISO) [1], International Electrotechnical Commission (IEC) [2], European Committee for Standardization (CEN) [3], and European Committee for Electrotechnical Standardization (CENELEC) [4] with the so-called "closed" approach require approval with "one-country-one-vote" from the country delegates, who are the only members and country representatives. The "closed" participation usually means participation in the closed society: for example, it can be the case of European Union members or of big countries in international organizations. Usually, the standardization work is done by national delegation, representing standardization organization of a certain country. Other international standardization organizations rely on the "open" approach for standardization process (examples are International Telecommunication Union (ITU) [5] and European Telecommunications Standards Institute (ETSI) [6]). In the "open" approach, delegates participate "directly," meaning that they form at the same time the decision body, deciding on what to standardize, on timing and on task resourcing, and on the final draft approvals. Most of the organizations, independently of the "closed" or the "open" approach, decide on the standard approval by consensus.

In conclusion, the technical standards life cycle of standards documents, including approval process, depends on the organization, more exactly on the membership of the certain standardization organization. The membership decides on the status of the documents that it produces. For example, approval can be given by the Technical Committee for Publicly Available Specification (PAS), Technical Specification (TS), or Technical Report (TR), but only the full membership can give approval for the International Standard (IS) and the European Standard (EN). As it will be shown in the further text, similar processes are used in the global and transnational standards bodies with the same type of approach to standardization and approval process. However, some global organizations produce documents that transnational organizations do not produce (e.g., CEN [3] and CENELEC [4] do not publish Publicly Available Specification).

5.2.1 *International Standards Life Cycle*

ISO

The development of standards depends on the standardization organization and its internal processes. In ISO, it takes 18–30 months to develop the standard from a New Work Item Proposal (NWIP) to a publication of an International Standard (IS).

It takes another 5 months for a review and ballot by all ISO members. The IS is approved if more than 67% of "P-Membership"[1] of the Technical Committee responsible for the document are for it and no more than 25% of the total votes are against. In the counting, negative not elaborated technically reasonable comments and abstentions are not counted. Another 3 months are required to review the resolution of comments. Ballot by TC members may take 2 or 4 months by agreement with the TC. The Final Draft International Standard (FDIS in approval stage), ISO Publicly Available Specification (ISO/PAS), and Technical Report (TR) are published by approval of at least a simple voting majority. Technical Specification (TS) is published by approval of at least 67% of the voters.

As shown in Fig. 5.4 and Table 5.1, the route consists of six phases, also presented in Table 5.2. The first phase (1) starts from defining the New Work Item Proposal (NWIP). The second phase (2) is about building expert consensus with deliverables of the first Committee Draft (CD) of International Standard or ISO Publicly Available Specification. In the third phase (3), consensus has to be built within TC/SC, with deliverables of the Draft International Standard (DIS) or a Technical Specification. It can be also a Technical Report, as a nonnormative document. The fourth phase (4) is based on the DIS Enquiry, with the deliverable of the Final DIS. In the fifth phase (5), the DIS Formal Vote follows with the deliverable of the Final International Standard (FIS). Finally, ISO IS is the deliverable of the last sixth phase (6) that deals with ISO IS Publication.

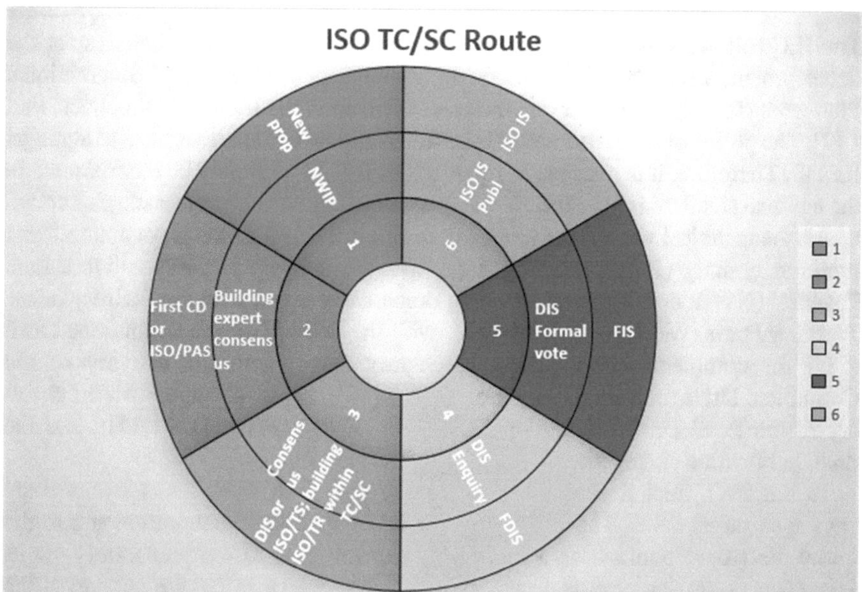

Fig. 5.4 ISO TC/SC route to the new ISO IS

[1] P-Member is a Participating Member. As it will be explained in Chap. 6, P-Members are active contributors in the meetings and have the obligation to vote at all stages of standards development.

Table 5.1 ISO TC/SC route

ISO TC/SC route phases		Deliverables
1	NWIP	New proposal definition
2	Building expert consensus	First CD or ISO/PAS
3	Consensus building within TC/SC	DIS or ISO/TS; ISO/TR
4	DIS Enquiry	FDIS
5	DIS Formal Vote	FIS
6	ISO IS Publication	ISO IS

Table 5.2 IEC TC/SC route

IEC TC/SC route stages		Deliverables
1	Preliminary stage (PWI)	
2	Proposal stage	NWIP, NP
3	Preparatory stage	WD
4	Committee stage	CD
5	Enquiry stage	CDV
6	Approval stage	FDIS
7	Publication stage	IEC IS

IEC

The IEC follows similar standards development processes and procedures for the development, drafting, and subsequent maintenance of the IEC International Standards (IEC IS) and other deliverables, as given in the ISO/IEC Directives, Part 1 [7]. The policy of IEC requires that the IEC National Committees have to agree on IEC IS. Therefore, it is possible to refer to it as IEC IS, or it may be incorporated in the national standards of the IEC community. The sequence of project stages through which the technical work is developed is shown in Table 5.2, and it goes as follows: preliminary stage (PWI); proposal stage, with the delivery of the New Work Item Proposal (NWIP or NP) (FormNP; [8]); preparatory stage, with the delivery of the Working Draft (WD); committee stage, with the delivery of the Committee Draft (CD) for comments (FormCD; [9]); enquiry stage, with the delivery of the Committee Draft for Vote (CDV) (FormCDV; [10]); approval stage, with the delivery of the Final Draft International Standard (FDIS) (FormFDIS; [11]); and the final, publication stage, with the delivery of the IEC IS.

In the PWI, Technical Committees (TCs) deal with work items without any envisaged target dates. The work is dedicated to the NWIP elaboration and to the initial draft development, as a requirement to progress to the preparatory stage. Some data collection or tests necessary to develop standards that are not part of the usual standardization process could also be performed in this stage. The proposal stage consists of taking the proposal for a new work that is, in the general case, made by industry, but it may also be made by a national body, TC/SC Secretariat, another TC/SC, a liaison organization, the Technical Management Board or one of its advisory groups, or the Chief Executive Officer.

The National Committee presents the proposal. A NWIP (or NP) is a proposal for the new standard, or it can be a new part of an existing standard or a Technical Specification (TS). TC/SC members usually receive a form from the proposer.

This stage results with the NP if 67% of the Committee's P-Members approve the new work item and if a minimum number of experts are nominated by P-Members approving the new work item proposal. In the case of 16 or less P-Members in the committees, a minimum of 4 experts and, for committees with 17 or more P-Members, a minimum of 5 experts should approve the NP.

In the preparatory stage, a project leader of a project team prepares a Working Draft (WD). If WD is not supplied with the proposal, its availability is 6 months. When WD is sent to the TC/SC members as the first Committee Draft (CD) and when registered by the CEO, the preparatory stage ends. In the case of special market needs, this phase can be also the final phase for publishing the final WD as a PAS.

In the committee stage, National Committees (NCs) receive CD for the comment. In the next 12 months, NCs study the CD and submit all comments. All P- and O-Members of the TC/SC have a typical deadline of 8, 12, or 16 weeks to submit the reply. It is important to reach consensus on the technical content.

The enquiry stage gives all NCs 12-week voting period as a possibility to submit their last comments on the bilingual Committee Draft for Vote (CDV). At least two thirds of the P-votes has to be in favor for the proposal, and less than one quarter of all the votes cast has to be negative about it, in order for the CDV to be approved. The Central Office Secretary sends a revised version within 16 weeks for the Final Draft International Standard (FDIS) processing. If there are no negative votes and no technical changes, CDV automatically proceeds to publication. For publishing the Technical Specification (not an International Standard), only two thirds of the votes must be positive about it. If so, the revised version is automatically sent to the Central Office for publishing.

In the approval stage, the National Committees receive Final Draft International Standard (FDIS) for the 6-week voting period. The NC's vote must be explicit: positive (without any comments), negative, or abstention. If 67% of P-Members are positive and less than 25% of all votes submitted are negative, FDIS is approved and allowed to the final, publication stage within 90 days with only minor editorial changes in the final text. If voting for FDIS is negative, FDIS goes back to the TC/SC for the reconsideration.

In the publication stage, the Central Office takes care of publishing International Standard within 6 weeks of FDIS approval.

Figure 5.5 shows stages of TC/SC IEC IS development.

A TC/SC defines its program considering sectoral planning requirements related to new projects and published IS maintenance within its agreed scope. IS Requests can be initiated also from the TC external environment, i.e., from the other Technical Committees, advisory groups of the Technical Management Board, policy-level committees, and organizations outside ISO and IEC. Each project has a number, which it will retain until its end. The number may be subdivided if the project itself

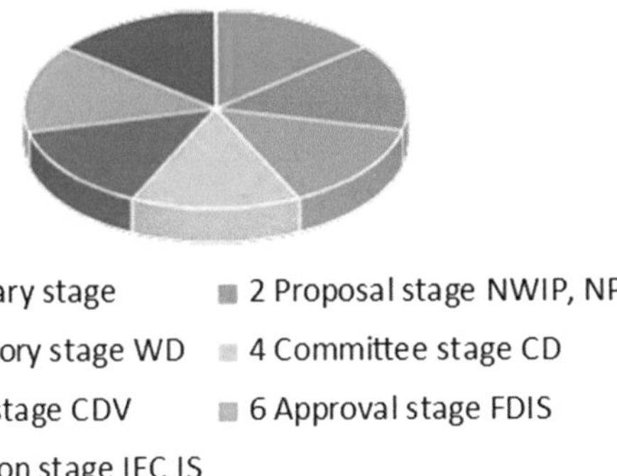

Fig. 5.5 TC/SC IEC IS development

should be subdivided. However, a NWIP has to be opened if the subdivision does not lie fully within the original scope. The work program indicates allocated subcommittees and/or working groups or project teams.

For an effective standards development, TC/SC establishes target dates for completion of the first Working Draft (usually 6 months), for the first Committee Draft circulation (12 months), for the Enquiry Draft circulation (24 months), for the Final Draft International Standard circulation (33 months), and for the International Standard publication (36 months). The defined target dates are clearly written in the work program, and they are under a continuous review. All work items are cancelled if they last longer than 5 years in the work program and if they are not reaching the approval stage.

Table 5.3 shows stage codes of IEC IS.

The TS is approved in a similar way as a standard, but the final vote goes with a simple majority of TC/SC Participating Members at Draft Technical Specification (DTS) stage, immediately after the Committee Draft (CD) stage. IEC TR is approved by TC/SC in the same way. The member bodies receive the IEC Draft Guide, adopted by the TC or Group, for voting. At least 75% of the member bodies casting a vote approve the Guide.

After the description of the IEC standardization process, the description of the IEC publication life cycle follows.

A development of a new publication or of an update or of a maintenance of an existing publication can take place in the preparatory and committee stage with WD and CD documents.

A new publication development starts with the definition of the exact text (Word) format, to avoid any kind of compatibility issue. After creation of a new Word document, the latest version of the IEC template has to be downloaded [12] and attached to the Word document, by consulting ISO/IEC Directives, Part 2 [13].

An update of an existing publication starts by requesting the revisable files [14] of the published version of the previous edition in the form of MS Word file and editable images from the IEC. The version from the FDIS stage of the previous

Table 5.3 IEC IS stage codes

Code	Meaning	Deliverable
ACD	Approved for CD	IS, TS, TR
ACDV	Approved for CDV	IS, TS, TR
ADTR	Approved for DTR	TR
ADTS	Approved for DTS	TS
AFDIS	Approved for FDIS	IS
APUB	Approved for publication	IS, TS, TR, PAS, ISH
BPUB	Being published	IS, TS, TR, PAS, ISH
CAN	Draft cancelled	IS, TS, TR, PAS, ISH
CD	Draft circulated as CD	IS, TS, TR
CDM	CD to be discussed at meeting	IS, TS, TR
CCDV	Draft circulated as CDV	IS
CDISH	Draft circulated as DISH	ISH
CDVM	Rejected CDV to be discussed at a meeting	IS
CFDIS	Draft circulated as FDIS	IS
CDPAS	Draft circulated as DPAS	PAS
CDTR	Draft circulated as DTR	TR
CDTS	Draft circulated as DTS	TS
DTRM	Rejected DTR to be discussed at meeting	TR
DTSM	Rejected DTS to be discussed at meeting	TS
DECDISH	DISH at editing check	ISH
DECFDIS	FDIS at editing check	IS
DECPUB	Publication at editing check	IS, TS, TR, PAS
DEL	Deleted/abandoned	IS, TS, TR
DELPUB	Deleted publication	IS, TS, TR, PAS, ISH
NCDV	CDV rejected	IS
NDTR	DTR rejected	TR
NDTS	DTS rejected	TS
NFDIS	FDIS rejected	Is
PCC	Preparation of CC	IS, TS, TR
PNW	New work item proposal	IS, TS
PPUB	Publication issued	IS, TS, TR, PAS, ISH
PRVC	Preparation of RVC	IS
PRVDISH	Preparation of RVDISH	ISH
PRVD	Preparation of RVD	IS
PRVDPAS	Preparation of RVDPAS	PAS
PRVDTR	Preparation of RVDTR	TR
PRVDTS	Preparation of RVDTS	TS
PRVN	Preparation of RVN	IS, TS
PWI	Preliminary work item	IS, TS, TR
RDISH	DISH received and registered	ISH
RFDIS	FDIS received and registered	IS
RPUB	Publication received and registered	IS, TS, TR, PAS

(continued)

Table 5.3 (continued)

Code	Meaning	Deliverable
TCDV	Translation of CDV	IS
TDISH	Translation of DISH	ISH
TDTR	Translation of DTR	TR
TDTS	Translation of DTS	TS
TFDIS	Translation of FDIS	IS
TPUB	Translation of publication	IS, TS, TR
WPUB	Publication withdrawn	IS, TS, TR, PAS, ISH

edition should not to be used. The latest version of the IEC template should be downloaded [12] and attached to the Word document. The foreword has to provide a document history, giving a brief list of changes with respect to the previous edition.

In the preparation process of a new publication and/or update, different authors contribute to one document. It is very important to keep control of document versions and handle only one central document, so that the efforts are not split in development of two or more parallel documents in the process of merging contributions that can later lead to the need of unnecessary repeating work and time consumption. Therefore, in times before the existence of the cloud, it was recommended to ensure that all the active contributors know exactly where is the original file. Of course, nowadays, it is much easier to do all the corrections in the same file in the cloud.

In the enquiry stage, a document reaches the CDV stage with the start of an involvement of the IEC Editing and Document preparation team in the IEC publication development, during the voting period of 12 weeks. The IEC Editing and Document preparation team is made of an IEC-internal team of editors, graphic designers, and layout specialists, whose role is to assist the TCs, especially to collaborate with the TC Secretary to facilitate preparation of the IEC publication. Since a complete set of CDV documents usually consists of the Word document plus image files [15] in editable formats for all figures, the IEC editors perform changes and comments in MS Word revision mode. The edited document is sent from the IEC to the TC secretariat and a copy to the French NC, because of the bilingual rule. The TC Secretary integrates the comments from the NCs after the CDV vote into the document and reviews the changes and comments from the IEC editors. After the period when all changes are performed, the document is ready for the next stage, with the approval of FDIS document. For a smooth arrival to the next phase of IEC publication development, it is necessary that any parallel development of the document is avoided, meaning that the TC does not make any change to the document, during the period that starts by sending CDV to the IEC, and ending by getting it back from the IEC, keeping control of document versions [16]. In this stage, the documents are exchanged via e-mail. If their size is bigger than 5 MB, the TC's Collaboration Space can be used.

During the CDV stage, it is important that the TC does not work on its copy of the document while the IEC is working on its editing.

The approval stage consists of the IEC receiving a draft FDIS, thorough IEC entry control, with examination of layout, tables, and texts, i.e., checking the usability of electronic file. It consists of using the IEC template; figures, i.e., their usability; and editing, i.e., did the CDV editing come into the document. The other set of questions relates a compliance of the document to the ISO/IEC Directives, Part 2, and a completeness of translations, without any missing texts. If all these questions are correctly answered, the IEC entry control gives "green light" to editing acceptance. Upon acceptance, the documents are the priority and they are ready within a period of 45 days. FDIS documents are ready within 90 days, according to the ISO/IEC Directives, Part 1 [7]. The documents undergo formatting and editing; the figures are reviewed and, if necessary, modified. The TC Secretary has the obligation to reply within 10 working days to any question. After approval of the IEC Technical Officer, the document is circulated for the vote. After approval of the publication, as the result of FDIS vote, the document should not be changed for any technical detail, but only for obvious editorial mistakes.

The IEC Central Office keeps published standards in editable file format (Microsoft Word). It provides them either to the relevant TC to update an existing standard or to the NC to prepare a national adoption of the requested IEC standard(s). All files are copyrights. Thus, they may only be used for the requested purpose. IEC Central Office can give specific authorization for any other requested use.

Maintenance and/or update of a publication starts with the TC taking the publication files of the previous edition, and not intermediate versions in their possession. The revisable files in the editable form of Word version (if existing in the case of older files) and, if existing, editable images can be got through the relevant IEC Technical Officer [14]. The obtained files may be used only for maintenance/update.

NCs take care of national adoptions of IEC publication by downloading an editable copy from the IEC Revisable files database [17]. If more NC members need to have access to revisable files, the NC Secretary has to indicate the IEC username(s) for granted permission to download revisable files to the IEC Technical Information and Support Services. NCs have to indicate the corresponding national reference with the year of adoption and degree of correspondence (IDT or MOD) for each downloaded file, i.e., IEC reference within 6 months of the date of download.

The IEC template (iecstd.dot) contains all the styles and automation features for IEC publications. It is the basic layout of all IEC publications. IEV template (IEV.dot [18]) is used for IEV documents (IEC 60050). The IEC template can be used after authors save it once on their computers and then attach to every working document. The process consists of downloading the file "iecstd.dot" and saving it in Word's "Templates" folder on user's computer, attaching the template [19] to the working document and using the template's styles and features in MS Word. IEC templates can be found in Table 5.4.

User guide to the IEC template (IEC-template-V5_User-guide.pdf) can be downloaded from [20].

The IEC template file (iecstd.dot) can be downloaded from [12]. The user is not advised to open the template file when the browser asks for it. If it happens that the browser opens the file without asking whether to open or save it, it should be closed,

Table 5.4 IEC templates

Description	File name
IEC template, Version 5, in English	iecstd.dot
IEC template in French, Version 5	iecstd_f.dot
User guide to the IEC template (current version: 1.4)	IEC-template-V5_User-guide.pdf
Keyboard shortcuts for some frequently used Word commands	Useful-Word-shortcuts_v3.pdf
Administrative circular on Version 5 of the IEC template	201325e.pdf

because in this case a new document based on the template is saved, but the template itself is required. Therefore, the saving is performed by right-clicking on the downloadable file and selecting *Save target as….* The template should be saved in Word's *Templates* folder on the computer with a typical path to the *Templates* folder *C:\Users\[user name]\AppData\Roaming\Microsoft\Templates.*

The new document based on the template can be created. In Word 2010 the process goes as *File > New > My templates*, with a double-click on *iecstd.dot*. In Word 2007, it is similar to the process in Word 2010, but instead of *File*, a click the MS Office button in the upper left-hand corner should follow. The new document opens with the IEC standard template and the IEC styles, as well as automation features in the *Add-Ins* tab. The document should be saved as a *.docx* file.

If the IEC template is attached to the existing document, the document can be opened. In Word 2010 the process goes as follows: *File > Options > Add-Ins*; at the bottom, next to *Manage*, select *Templates*, and then click *Go*. In Word 2007, similar to Word 2010, but instead of *File*, one has to click on the MS Office button in the upper left-hand corner and then to select *Word options*. Figure 5.6 shows the window with opened *Templates and Add-Ins.*

The *Attach* (1) should be clicked and browsed to the location of *iecstd.dot*; *in the next step, the OK* should be clicked after.

The option *Automatically update document styles* (2) replaces all the styles in the existing document with the required IEC original styles. With the checked option, updating is on during the opening of the document. This first-time attachment of the template has to be checked, just to be sure that the template and corresponding styles are correctly incorporated. Once the styles are updated, the recommendation is to uncheck the option of automatic update, just to stop unwanted reset of all the numbers to automatic numbering of clauses and subclauses.

The next IEC recommendation concerns saving *Templates* icon to *Quick Access Toolbar* by right-clicking on the arrow to the right of the toolbar, selecting *Customize Quick Access Toolbar > More Commands*, selecting under *All Commands Choose commands*, and selecting *Templates* and clicking *Add* and then *OK* in the list of commands.

Drafting standards in the IEC 60050 series starts by downloading IEV standard template. The International Electrotechnical Vocabulary (IEV), IEC 60050 series of standards, is maintained by IEC TC 1. The IEV online is known as the Electropedia [21].

Fig. 5.6 The *Templates and Add-Ins* window

Table 5.5 IEV templates

Description	File name
IEV template, Version 7.2, in English and French	IEV.dot
IEV template styles	IEV.dot styles.pdf

The template *IEV.dot* can be downloaded and saved in Word's *Templates* folder on the computer. The next step is to attach the template to the working document and to use the template's styles and features in MS Word. Table 5.5 shows templates in English and French and template styles.

The cooperation between IEC and IEEE is given in Guide to IEC/IEEE cooperation [22]. The joint **IEC/IEEE publications** are written when there is a need to optimize IEC and IEEE resources in the development of standards. The joint development also shortens time to the global market. The joint IEC/IEEE publications use a special foreword, header, and footer that are provided in Table 5.6.

The main image editing tools for IEC standards are given in Table 5.7. They are Microsoft Word, Microsoft Excel, Microsoft PowerPoint, Microsoft Visio, Autodesk AutoCAD, Adobe Illustrator, Adobe Photoshop, Corel Designer, CorelDraw, Inkscape, and OpenOffice Draw.

Table 5.8 gives standards used in the creation of graphical content of IEC documents. General standard for preparation of documents used in electrotechnology is

Table 5.6 IEC/IEEE Standard text

Description	Document
Foreword (English)	IEC-IEEE_Harmonized-foreword_E.doc
Foreword (French)	IEC-IEEE_Harmonized-foreword_F.doc
Headers and footers	IEC-IEEE_Headers-and-footers.doc

Table 5.7 Main image editing tools

File extension	Software application
.doc, .docx	Microsoft Word
.xls, .xlsx	Microsoft Excel
.ppt, .pptx	Microsoft PowerPoint
.vsd	Microsoft Visio
.dwg	Autodesk AutoCAD
.ai	Adobe Illustrator
.psd, .tif, .jpeg, .png	Adobe Photoshop
.des	Corel Designer
.svg	E.g., Inkscape, Adobe Illustrator, CorelDraw
.odg	OpenOffice Draw

Table 5.8 Standards used in the creation of graphical content of IEC documents

Content	Standard	Title
General	IEC 61082-1	Preparation of documents used in electrotechnology – Part 1: Rules
Graphical symbols	IEC 60417	Graphical symbols for use on equipment
	ISO 7000	Graphical symbols for use on equipment – registered symbols
	IEC 62648	Graphical symbols for use on equipment – guidelines for the inclusion of graphical symbols in IEC publications
	IEC 60617	Graphical symbols for diagrams
	IEC 80416-1	Basic principles for graphical symbols for use on equipment – part 1: Creation of graphical symbols for registration
	ISO/IEC 81714-1	Design of graphical symbols for use in the technical documentation of products – Part 1: Basic rules
Line types	ISO 128-20	Technical drawings – General principles of presentation – Part 20: Basic conventions for lines
Dimensioning	ISO 129 all parts	Technical drawings – Indication of dimensions and tolerances
Dimensional and geometrical product specifications	ISO 1101	Geometrical product specifications (GPS) – Geometrical tolerancing – Tolerances of form, orientation, location and run-out
Projection	ISO 128-30	Technical drawings – General principles of presentation – Part 30: Basic conventions for views
Flowcharts or organigrams	ISO 5807	Information processing – Documentation symbols and conventions for data, program and system flowcharts, program network charts and system resources charts

IEC 61082-1; the standard for graphical symbols is IEC 60417 for use on equipment; ISO 7000 contains registered symbols for use on equipment; IEC 62648 contains guidelines for the inclusion of graphical symbols in IEC publications; IEC 60617 contains graphical symbols for diagrams; IEC 80416-1 contains creation of graphical symbols for registration; and ISO/IEC 81714-1 contains basic rules for the design of graphical symbols for use in the technical documentation of products. The standard ISO 128-20 deals with the basic conventions for lines in technical drawings. All parts of ISO 129 deal with the indication of dimensions and tolerances in technical drawings. The standard ISO 1101 explains geometrical tolerancing of form, orientation, location, and runout in geometrical product specifications. The standard ISO 128-30 deals with the basic conventions for views of technical drawings within the general principles of presentation. The standard ISO 5807 presents the documentation symbols and conventions for data, program and system flowcharts, program network charts, and system resources charts in information processing.

ITU-T

In Chap. 6 organization of the ITU is explained in more detail. This chapter discusses the life cycle of the ITU-T documents. The part of ITU that specifically deals with standards is the ITU Telecommunication Standardization Sector (ITU-T) [23]. The ITU-T membership contributes with inputs to a Study Group (SG), suggesting typically new work areas, Draft Recommendation, or changes to the existing Recommendations. This is the basis of a contribution-driven environment in the ITU-T. However, The ITU-T is also a consensus-based environment.

The basic project unit within ITU-T is a Question, whose text defines the area of a study project. A new Question is established when a number of Members commit to support the work, meaning that the Study Group (SG) approves it. The work of SGs is in the form of study Questions, and it is led by a Chairperson and several vice-chairs appointed by the World Telecommunication Standardization Assembly (WTSA). Questions are driven by contributions, requesting technical studies in a specific area of the telecommunication standardization. They are terminated after the completion of the defined work. The work continues if there are more technical, market-oriented, network or service-driven developments. Texts for all the Questions are on the web page of the specific SG. The SG can consist of Working Parties (WPs), which coordinate study Questions on a certain related theme. A Rapporteur group is a team of experts, who work on a specific Question. The participants determine Recommendations to be developed and develop text for them by consulting other relevant parts of ITU-T. Experts meet to progress the work during parent SG/WG, but it is also possible to have meetings in a more informal way.

Traditional Approval Process (TAP) is used for standards with regulatory implications. The electronic handling of documents without any physical meetings continues after the beginning of the approval process until the end of preparation of a Draft Recommendation by the SG experts.

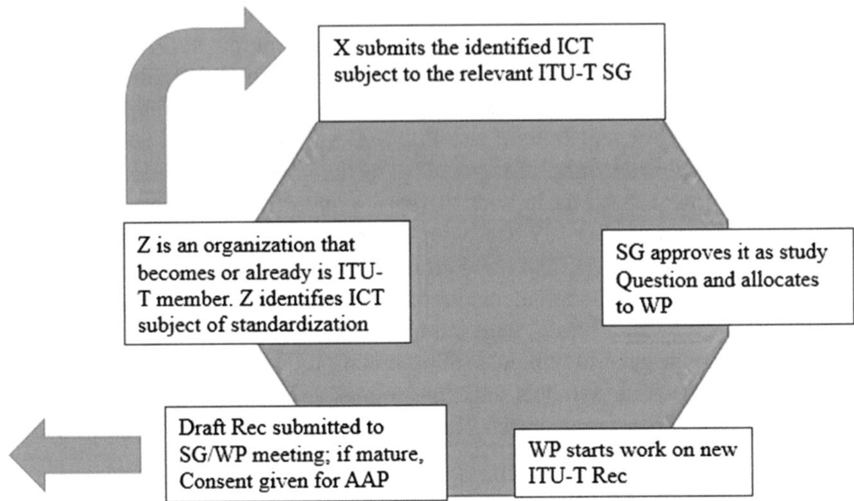

Fig. 5.7 The first part of the ITU-T AAP

Standards may be brought to the market in a very fast manner, employing a process that has been available since 2001, entitled "Alternative Approval Process" (AAP). The AAP is shown in Fig. 5.7.

A standard now needs 2 months, or even 5 weeks for approval, in comparison with the earlier 4 years (in the mid-1990s) or 2 years (until 1997). AAP had the chance to appear due to the electronic handling of documents and global interconnections that did not exist earlier. AAP is important in strengthening public/private partnership (PPP), because the both public (Member States) and private members (Sector Members) participate in the approval process.

During the first part of the ITU-T AAP, let us take an example of the organization named "Z." Z can be already a member organization of the ITU-T, or it is just becoming a member. In any case, Z identified the specific ICT subject of standardization. X then submits the identified ICT subject to the relevant ITU-T SG. If the members agree, the SG can approve it as the study Question. Then, the SG allocates it to the WP. Based on that the WP now starts the work on the new ITU-T recommendation. As the final phase of the first part of the ITU-T AAP, the Draft Recommendation is submitted to the SG/WP meeting. If the Draft Recommendation is mature, the Consent is given for the AAP.

During the second part of the ITU-T AAP (shown in Fig. 5.8), SG or WP meeting reviews the Draft Recommendation. If there is an agreement during the meeting, a Consent is given, as an initiation point leading to the final review of the Draft Recommendation. At that point of time, the Director of ITU-T's Secretariat, Telecommunication Standardization Bureau (TSB), starts the approval period and the so-called Last Call Phase (LCP) of the AAP procedure, lasting for 4 weeks. The draft text is posted on the ITU-T website [23]. All Member States and Sector Members can review and comment the text.

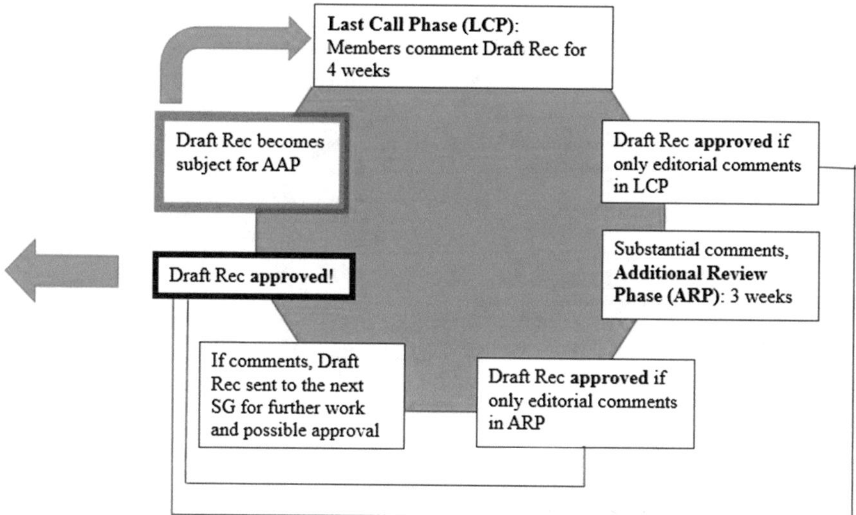

Fig. 5.8 The second part of ITU-T AAP

If the received comments are of editorial nature only, the Recommendation is considered approved. In the case of receiving any comments in relation with the substance of the text, SG Chair initiates the comment resolution process in the joint action with TSB and the concerned experts. This period of "Additional Review Phase" (ARP) lasts for 3 weeks, and it ends with the Additional Review of the posted revised text on the website. Again, if received comments are of only editorial nature, the Recommendation is considered approved. If there are still issues with the document, the next SG meeting should further discuss the draft text and all received comments and work toward gaining possible resolution and approval.

5.2.2 *Supranational Standards Life Cycle: EU*

The "New Approach" system of the European Union (EU) defines three recognized European Standards Organizations (ESOs) that can publish the official EU standards: CEN [3], CENELEC [4], and ETSI [6]. All three of them will be discussed in Chap. 7 on supranational innovation and standards circles with an example of the European Union. In the continuation of this chapter, the life cycle of all three ESOs will be presented.

CEN

CEN [3] stands for the European Committee for Standardization. CEN is a non-profit technical organization, and it was founded according to the Belgian law in 1961. CEN is the only confirmed European organization for planning, writing, and

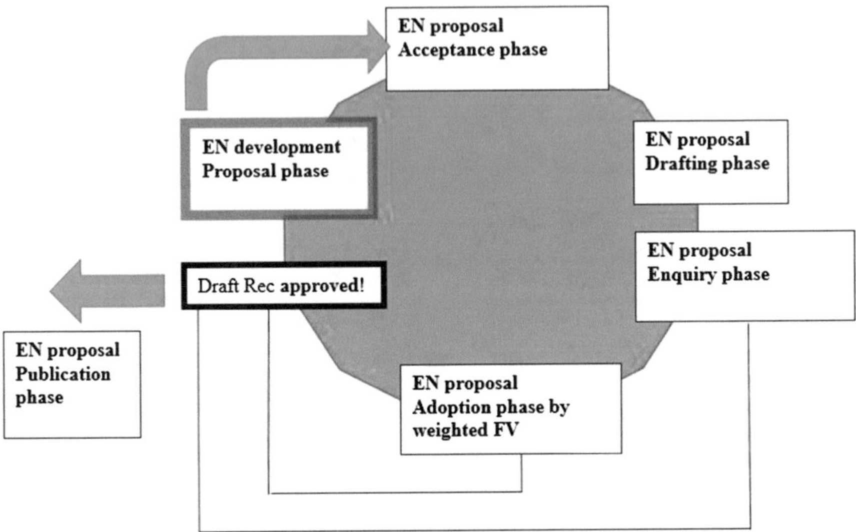

Fig. 5.9 Summary of CEN EN process

adopting the European standards in all areas of economic activity except in electrical engineering. The development of CEN European Standard (EN) defines high principles of openness, transparency, national commitment, technical coherence, and consensus. All these principles are defined in the CEN's Business Operations Support System (BOSS) [24]. BOSS provides practical help in terms of information for the development of deliverables within the CEN system. This means that the active participant of CEN can find on the BOSS pages clear description of the processes and instructions for the daily work with CEN documents.

The summary of the CEN EN process is illustrated in Fig. 5.9.

It starts with a proposal phase. Anybody can submit a proposal, including CEN TC or SC, European Associations, and EU/EFTA, which means that all the CEN members are welcome to suggest a subject for CEN standardization. However, most of the standardization work is proposed by, or through, the CEN members. In the next CEN EN proposal phase, acceptance of the proposal, the Technical Body or Technical Board has to accept the EN proposal. At that point of time, the "standstill" period starts as an obligation for the Member Countries. "Standstill" encompasses holding all the national activities within the scope of the proposed EN, i.e., not starting any new projects nor revising existing national standards. In the third, Drafting phase, experts of a Technical Body develop CEN EN. During the Enquiry phase, which starts once the draft EN is ready, everybody with an interest in the topic has the possibility to comment it. The interested parties could be public authorities, consumers, or manufacturers. The comments are gathered and submitted by Member States as national public comments by weighted vote to CEN Technical Body, which decides to publish (or not) the CEN EN, depending on the EN approval. If the decision is not to publish draft CEN EN, relevant Technical Body must update the draft EN and resubmit it for another weighted vote (Formal Vote), giving this phase name: Adoption by weighted Formal Vote. When the draft CEN EN is approved (in Enquiry or Formal Vote), EN can be published. After the CEN EN publishing phase, all the Member States have the obligation to approve it as a national standard, meaning that they have to withdraw any conflicting national standards. This step is obligatory due to the enabling of easier access to the whole CEN Member States market when applying European Standards. Within 5 years of its publication, European Standard has to be reviewed, in order to estimate its relevance. In this, CEN EN review phase, EN can be confirmed, revised, modified, or withdrawn.

CENELEC

CENELEC [4] is the European organization (one of the three ESOs that can issue European Standard) that is responsible for the area of electrical engineering. The European Committee for Electrotechnical Standardization, CENELEC, is a nonprofit organization founded in 1973.

The standardization process of EN development of CENELEC is the same as of CEN. It means that high principles of openness, transparency, national commitment, technical coherence, and consensus, as defined in BOSS [24], are preserved also in the development of CENELEC European Standard (EN).

The summary of CENELEC EN process is very similar to the summary of CEN EN process.

The CENELEC EN development starts with the proposal phase. Anybody can give a proposal, including CENELEC TC or SC, European Associations, and EU/EFTA (as CEN). All the CENELEC Members are welcome to suggest a subject for CENELEC standardization. As in CEN, most of the standardization work is proposed by or through the CENELEC Members. In phase of acceptance of the proposal, the Technical Body or the Technical Board has to accept the EN proposal. The "standstill" period starts as an obligation for the Member States. "Standstill" means holding all the national activities within the scope of the proposed EN: Member States are not allowed to start any new projects nor to revise existing national standards. In the third drafting phase, experts of a Technical Body develop CENELEC EN. Starting with the ready draft EN, the Enquiry phase gives everybody (e.g., public authorities, consumers, or manufacturers) with an interest in the topic a possibility for comments. Member States gather comments and submit national public comments by weighted vote to the CENELEC Technical Body, which decides to publish (or not) the CENELEC EN, depending on the EN approval. Only in the case of 100% approval for the EN, the European Standard is published. If the decision is not to publish draft CENELEC EN, the relevant Technical Body has to update the draft EN and resubmit it for another weighted Formal Vote, giving this phase name: Adoption by weighted Formal Vote. When the draft CENELEC EN is approved (in Enquiry or Formal Vote), EN can be published. After the CENELEC EN publishing phase, all the Member States have the obligation to approve it as a national standard, meaning that they have to withdraw any conflicting national standards. This step is obligatory due to enabling easier access to the whole CENELEC Member States market when applying European Standards. Within 5 years of its publication, European Standard has to be reviewed, in order to estimate its relevance. In this, CENELEC EN review phase, EN can be confirmed, revised, modified, or withdrawn.

European Standards are published in three official languages: English, French, and German. It is allowed for National Committees to translate standards in their own language. Numbering of a European Standard consists of the capital letters EN followed by a space and a number in Arabic numerals, without any space. As an example, EN 50122-1: 2011 means that the document is the first part of the EN with number 50122 that was published in the year 2011. The first two numerals indicate the origin of the standard:

- 40000 to 44999 cover domains of common CEN/CENELEC activities in the ICT field.
- 45000 to 49999 cover domains of common CEN/CENELEC activities outside the ICT field.
- 50000 to 59999 cover CENELEC activities, i.e., standardization work undertaken purely at European level.
- 60000 to 69999 refer to the CENELEC implementation of IEC documents with or without changes.

The summary of the collaboration between CENELEC process is presented in Fig. 5.10. Proposal of the standard comes from CENELEC TC/SC, the European Association, or EU/EFTA to the Technical Office decision of BT. It can either be directed directly to the IEC or the new CENELEC project can be started. It starts with the Draft "prEN," after which the comments by the NCs are collected in the

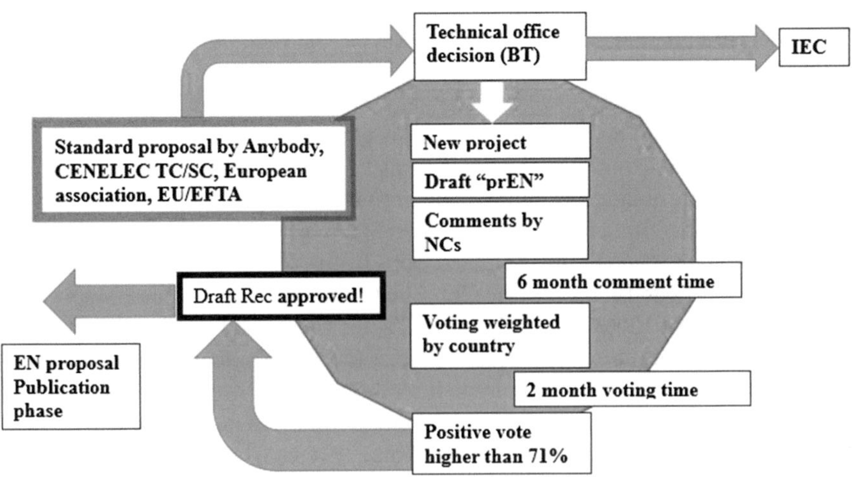

Fig. 5.10 Summary of CENELEC-IEC EN process

period of 6 months. After that period, voting is weighted by the countries in the period of the next 2 months. If the positive vote is higher than 71%, the Draft Recommendation is approved. In this case, the EN proposal publication phase starts.

ETSI

The European Telecommunications Standards Institute (ETSI) [6] is the third confirmed European organization and ESO that can issue European standard. ETSI is responsible for telecommunications.

Standardization principles of ETSI are as follows:

(a) Openness
(b) Consensus
(c) Transparency
(d) Coherence
(e) Development
(f) Relevance
(g) Equity
(h) Effectiveness
(i) Impartiality

ETSI standards and Technical Specifications are based on the open decision-making that is transparent, collaborative and consensus based, and accessible to all interested parties. Transparency is shown in all archived and identified information related to the technical discussions and decision-making, related to new standardization activities, related to the participation of all relevant categories of interested parties for achieving balance, and related to consideration and response to comments by the interested parties.

Published ETSI standards and Technical Specifications are relevant, because they respond to the market needs and/or regulatory requirements. They have a long-period guarantee of ongoing support and maintenance. They can be implemented on defined terms (either free of charge or for a reasonable fee), since they are publicly available. The standardized interfaces and testing, as key parts, are the result of agreement of all the organizations that adopted ETSI standards and Technical Specifications. Both ETSI standards and technical specifications are high-quality documents that are based on the advanced scientific and technological developments that contain detailed levels for development of a variety of competing interoperable products and services and demonstrate improvements. It is an important value of standard to initiate implementers for developing a competition and an innovation. This is the reason for adoption of ETSI documents all over the world. ETSI is oriented toward conforming to the highly respected and balancing Intellectual Property Rights (IPR) policy [25], fulfilling the needs of standardization for public use with the IPR owners. Licensing of Intellectual Property Rights (IPR), defined in a timely manner, can be without monetary compensation but still on (fair) reasonable and nondiscriminatory terms ((F)RAND).

A new proposal for starting a work item, meaning the creation of a new standard or an update of an existing standard, has to be launched by at least four ETSI members and has to be agreed by the relevant standards group.

Technical Committees (TCs) have a leading role in writing most of ETSI standards. TC members are representatives of ETSI members. A "Rapporteur" is a leader of the action. ETSI members have the right for participation in any group and work activity, except of certain security-related work, with the controlled participation by the ETSI Board. Specialist Task Forces (STFs) are groups of technical experts working together intensively for a defined period on specific items, usually in the case of an urgent need. Industry Specification Groups (ISGs) work in specific technology areas and are an effective alternative to industry fora.

ETSI documents are approved either by the participants in the relevant committee or the entire ETSI membership. For European Standards, ETSI's National Standards Organizations (NSOs) have the highest power to approve the final version.

Technical Specifications (TS), Technical Reports (TR), Group Specifications (GS), Group Reports (GR), and Special Reports (SR) are delivered after TC or ISG create the work item, work during ETSI Deliverable Drafting Phase, approve it during ETSI Reference Body Approval, and submit it to the ETSI Secretariat for document publishing. Figure 5.11 shows the ETSI process.

The full ETSI membership approves ETSI Standards (ES) and ETSI Guides (EG) by "Membership Approval Procedure." ETSI Secretariat makes the document available to ETSI Members after approval of the draft by TC. All ETSI full and associate members vote on adoption of the standard. ETSI Secretariat publishes document if the vote was successful. If vote was not successful, the standard is referred to the TC. ETSI provides web-based approval mechanisms, contributing to openness and transparency of the process. ETSI approval process of ES and EG is shown in Fig. 5.12.

Approval of the European Standards (ENs) requires ETSI Secretariat to ensure that the document is available to NSO, after the TC approved the draft EN. The NSOs perform as a single-process public enquiry and a weighted national vote, with consultations and submission of the national position on the EN. If the vote is successful and consultation results do not require any significant changes, ETSI Secretariat finalizes the draft. As the next step, ETSI Secretariat publishes the standard. If there are technical comments received during public enquiry, the TC goes into draft revision and resubmission to the ETSI Secretariat. However, if the required

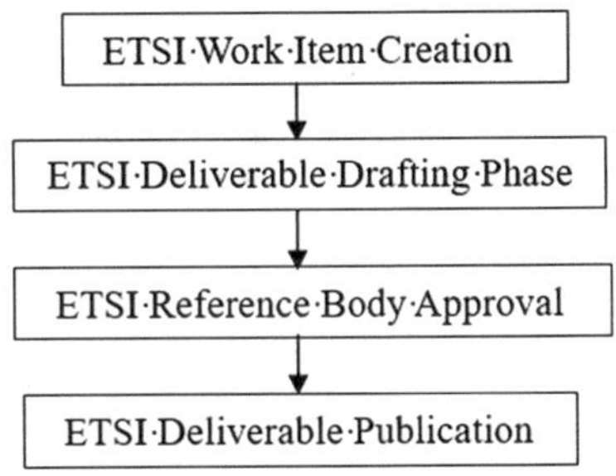

Fig. 5.11 ETSI approval process of TS, TR, GS, GR, and SR

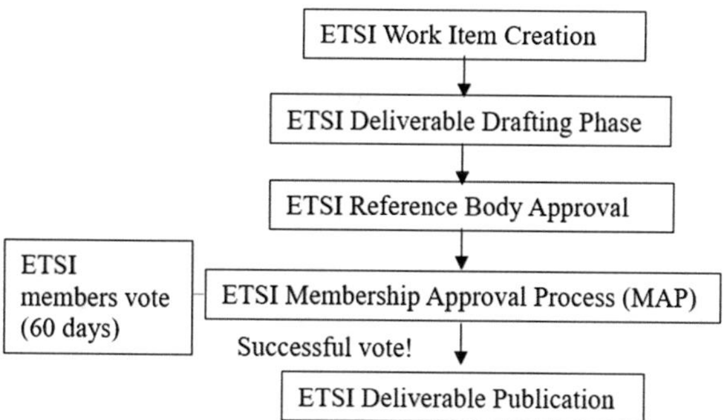

Fig. 5.12 ETSI approval process of ES and EG

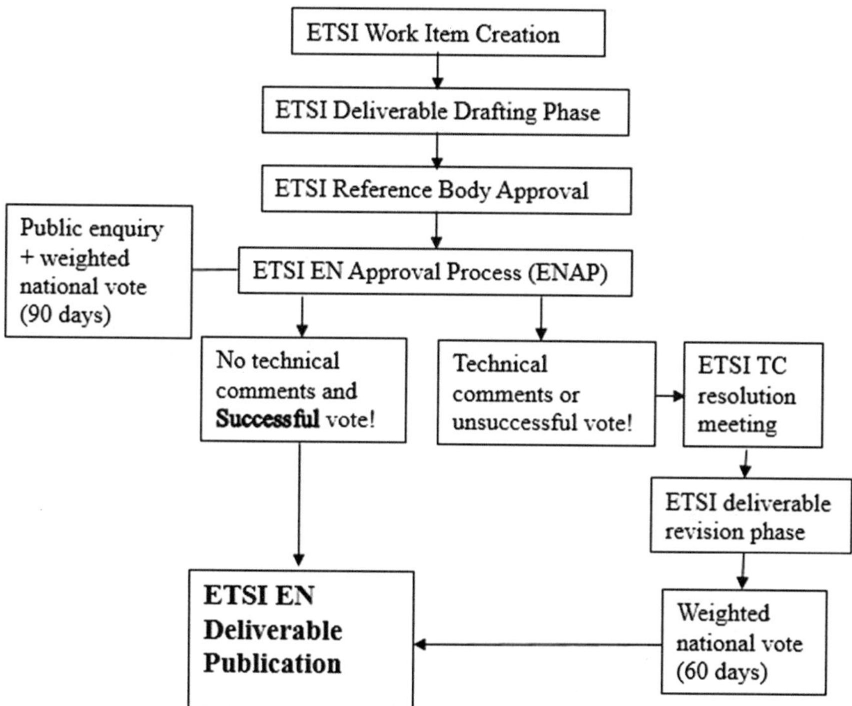

Fig. 5.13 ETSI approval process of EN

changes are significant, the Secretariat may initiate another public enquiry. The other option is that the draft is presented directly to a second vote. The approved standard is published by the ETSI Secretariat at ETSI headquarters. The Secretariat has the responsibility of ensuring that all the relevant procedures were completely followed. In the whole process, close cooperation rules between the Secretariat and the team that drafts the document. Only this kind of joint work can guarantee the high quality of the final document. In this case, ETSI Secretariat publishes EN (Fig. 5.13).

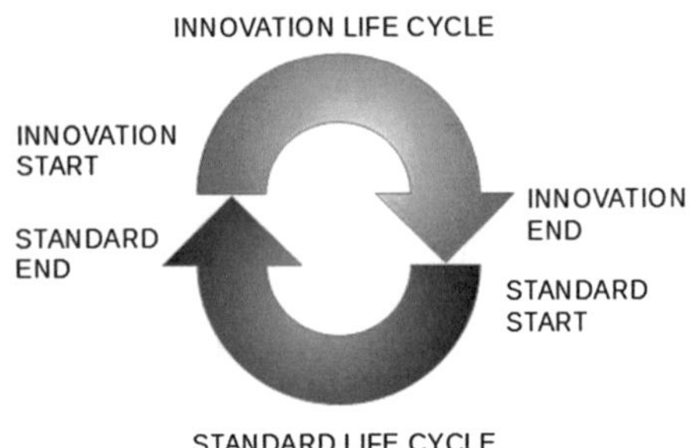

Fig. 5.14 Eternal life cycle of innovations and standards

All types of ETSI documents, except some Group Specifications, have successful vote if at least 71% of the weighted votes cast are positive about the draft. ETSI General Assembly decides on the weight of each nation's vote for European Standards. Weight of the vote of each ETSI member is agreed between the members for other types of ETSI documents. Fast evolving technology and developing market needs make ETSI working all the time on new documents and on documents maintenance and their update. ETSI Directives [26], and in particular the Technical Working Procedures, outline all the details of the approval process.

In conclusion, there is a huge interaction of innovations and standards, so that one without the other cannot exist. At the point of an end of a standard life (standard withdrawal), the innovation process starts again. In this sense, standards and innovation can be compared to a day and a night. At the point of the end of the innovation life, the standardization process starts. When day finishes, night starts, and vice versa. This is the correct explanation of the eternal innovation standards life cycle that is shown in Fig. 5.14.

Bibliography

1. ISO, Internet page: www.iso.ch. Retrieved 7.8.2019
2. IEC, Internet page: www.iec.ch. Retrieved 6.8.2019
3. CEN, Internet page: www.cen.eu. Retrieved 5.8.2019
4. CENELEC, Internet page: www.cenelec.eu. Retrieved 8.8.2019
5. ITU, Internet page: www.itu.int. Retrieved 7.8.2019
6. ETSI, Internet page: www.etsi.org. retrieved 6.8.2019
7. ISO/IEC Directives, Part 1, Internet page: www.iec.ch/standardsdev/resources/draftingpublications/directives/. Retrieved 20.8.2019
8. New Work Item Proposal (NP) (FormNP), Internet page: www.iec.ch/standardsdev/resources/forms_templates/documents/Form-NP.dotx. Retrieved 20.8.2019
9. Committee Draft for comments (CD) (FormCD), Internet page: www.iec.ch/standardsdev/resources/forms_templates/documents/Form-CD.dotx. Retrieved 20.8.2019
10. Committee Draft for Vote (CDV) (FormCDV), Internet page: www.iec.ch/standardsdev/resources/forms_templates/documents/Form-CDV.dotx. Retrieved 20.8.2019
11. Final Draft International Standard (FDIS) (FormFDIS), Internet page: www.iec.ch/standardsdev/resources/forms_templates/documents/Form-FDIS.dotx. Retrieved 20.8.2019

12. IEC template, Internet page: www.iec.ch/standardsdev/resources/draftingpublications/writing_formatting/IEC_template/iec_template.htm. Retrieved 20.8.2019
13. ISO/IEC Directives, Part 2, Internet page: www.iec.ch/standardsdev/resources/draftingpublications/directives/. Retrieved 20.8.2019
14. IEC publication maintenance, Internet page: www.iec.ch/standardsdev/resources/draftingpublications/overview/drafting_process/files_for_maintenance.htm. Retrieved 20.8.2019
15. IEC image files, Internet page: www.iec.ch/standardsdev/resources/draftingpublications/graphics_figures/exchanging_obtaining_figures/submitting_files_to_IEC.htm. retrieved 20.8.2019
16. IEC control version, Internet page: www.iec.ch/standardsdev/resources/draftingpublications/overview/drafting_process/control_doc_versions.htm. Retrieved 20.8.2019
17. IEC revisable files database, Internet page: http://std.iec.ch/iec-lib/revisablefile.nsf/welcome?readform. Retrieved 20.8.2019
18. IEV standard template, Internet page: www.iec.ch/standardsdev/resources/draftingpublications/writing_formatting/IEC_template/iev_standard_template.htm. Retrieved 20.8.2019
19. IEC writing formatting, Internet page: www.iec.ch/standardsdev/resources/draftingpublications/writing_formatting/IEC_template/attach_template_to_doc.htm. Retrieved 20.8.2019
20. IEC User guide, Internet page: www.iec.ch/standardsdev/resources/draftingpublications/writing_formatting/IEC_template/pdf/IEC-template_v5_User-guide.pdf. Retrieved 20.8.2019
21. Electropedia, Internet page: http://www.electropedia.org/. Retrieved 20.8.2019
22. Guide to IEC/IEEE cooperation, Internet page: www.iec.ch/iec-ieee. Retrieved 20.8.2019
23. ITU-T, Internet page: www.itu.int/en/ITU-T/Pages/default.aspx. Retrieved 21.8.2019
24. BOSS, Internet page: http://boss.cen.eu/developingdeliverables/Pages/default.aspx. Retrieved 20.8.2019
25. Intellectual Property Rights (IPR) policy, Internet page: www.etsi.org/intellectual-property-rights. Retrieved 20.8.2019
26. ETSI Directives, Internet page: portal.etsi.org/directives/40_directives_apr_2019.pdf. Retrieved 20.8.2019

Chapter 6
Global Innovation and Standards Circles

6.1 Global Innovation Circle

Global innovation ecosystem or Global Innovation Circle, GIC (shown in Fig. 6.1), consists of the same stakeholders as in Sect. 3.1 but with an organization on the global level. These are:

- Research and Education Institutions (REI) or Industry (IND) as a nurturing place of inventors
- Governmental and private organizations funding (FUN), forming the entrepreneurial level
- Marketing Organizations (MO), as organized places of marketers
- USers (US) and Users Organizations (UO) of a nongovernmental type

Globalization makes inventors, especially from the industry (IND), participating in the standardization process. In this case, the global innovation circle is shown in Fig. 6.2, with an added stakeholder (Global Standardization Organizations):

- Research and Education Institutions (REI) or Industry (IND) as a nurturing place of inventors, providing funding (FUN)
- Global Standardization Organizations (GSO)
- Industry (IND) producing invention
- Marketing Organizations (MO), as organized places of marketers
- USers (US) and Users Organizations (UO) of a nongovernmental type

Inventors can also be holders of Intellectual Property Right (IPR). In this case, the Global innovation circle has six stakeholders, as shown in Fig. 6.3:

- Research and Education Institutions (REI) or Industry (IND) (especially small and medium enterprises) as a nurturing place of inventors
- IPR organization, e.g., World Intellectual Property Organization (WIPO) [1]
- Global Standardization Organizations (GSO)
- Industry (IND) producing invention

D. Šimunić, I. Pavić, *Standards and Innovations in Information Technology and Communications*, https://doi.org/10.1007/978-3-030-44417-4_6

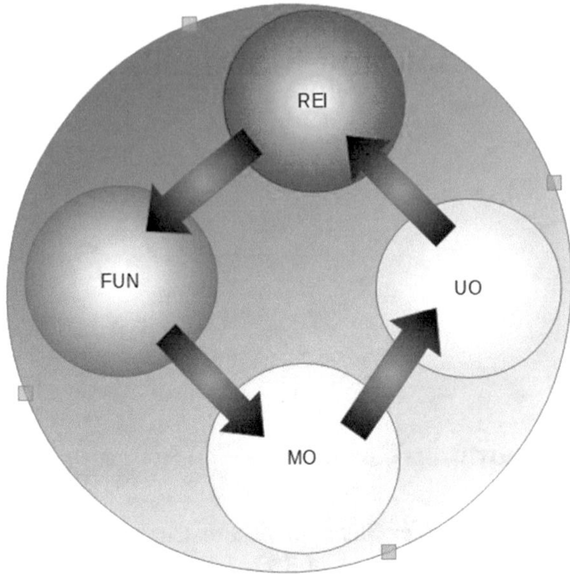

Fig. 6.1 Global innovation circle with the four stakeholders

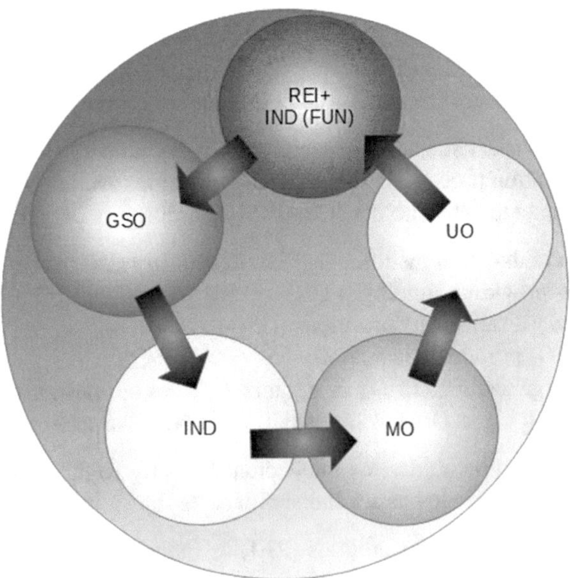

Fig. 6.2 Global innovation circle with the five stakeholders

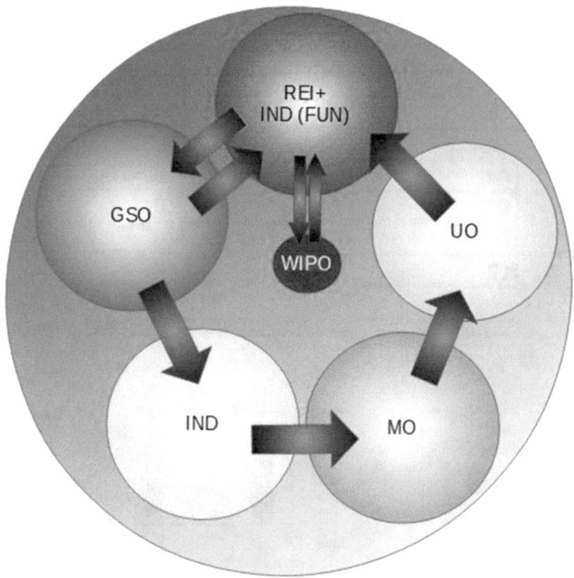

Fig. 6.3 Global innovation circle with the six stakeholders

- Marketing Organizations (MO), as organized places of marketers
- USers (US) and Users Organizations (UO) of a nongovernmental type

In the last case, the Global innovation circle has six stakeholders. Usually, it is a small and medium enterprise (SME) that is a REI with an IPR. In this case, REI first makes the invention, patenting it with the patent organization (e.g., global organization of that type is WIPO [1], or any other Global Intellectual Property Organization, GIPO). After obtaining the patent rights, REI promotes its invention in a GSO and contributes to the new prototype. Very often, the GSO either buys the rights itself. It can also happen that another company that has interest in participating in the valuation buys the patent rights. The rest of the global innovation circle is the same: after standardization, industry is producing the product, marketers are marketing it, and users will buy it and use it. Users will give their opinion to REIs and industry on how to improve or change it, and the Global innovation circle is closed.

There are many possible different configuration types of the Global innovation circle. The inspiration was taken from [2]. However, the most important are four cases, shown in Table 6.1. and in Fig. 6.4. The abbreviations related to the configuration types are the same as in Fig. 6.4, i.e., REI stands for Research and Education Institutions, FUN stands for Governmental and private organizations funding, MO stands for Marketing Organizations, UO stands for Users Organizations of a nongovernmental type, GSO stands for Global Standardization Organizations, and GIPO stands for Global Intellectual Property Organization. Furthermore, "GL" means "Global Level," "SNL" means "SupraNational Level," and "NL" means "National Level."

Table 6.1 The four important configuration types of global innovation circle

Stakeholder	GIC type			
	GOGIC	LMGIC	LKGIC	LGIC
REI	GL	GL	NL	NL
FUN	GL	GL	SNL	NL
MO	GL	NL	GL	NL
UO	GL	SNL	GL	NL
GSO	GL	GL	GL	GL
GIPO	GL	GL	GL	GL

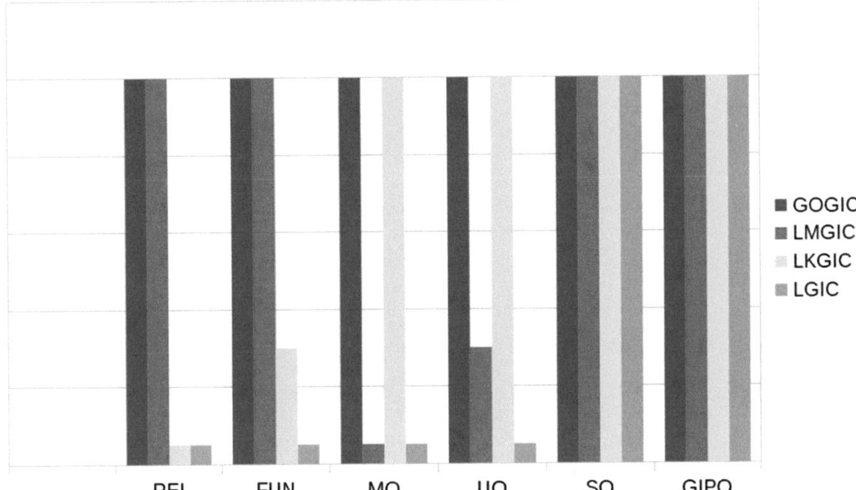

Fig. 6.4 Global innovation circle with the six stakeholders. NL means National Level, SNL means SupraNational Level, and GL means Global Level

As shown in Table 6.1 and Fig. 6.4, Globally Oriented GIC (GOGIC) is a full Global Innovation Circle, meaning that the knowledge is contained in international communities (on the Global Level, GL). Research and Education Institutions are global (GL), funding (FUN) of GOGIC is of the global type (GL), the market (MO) is oriented to the complete price competition (GL), the nature of users (UO) is global (GL), the standardization organization (GSO) is global (GL), and the intellectual property organization (GIPO) is also global (GL). An example of GOGIC could be consumer electronics. The second type, Local Market GIC (LMGIC), has the same characteristics of global knowledge and funding, as well as of global standardization and intellectual property organizations as the previous, Globally Oriented GIC, but the market (MO) has the local nature (i.e., it is on the National Level, NL), and the users (UO) have the supranational nature (i.e., SupraNational Level, SNL). The example could encompass carbon capture. The third type, Local Knowledge GIC (LKGIC), has a market (MO), users (UO), standardization (GSO) and intellectual property organizations (GIPO) of global nature, but funding (FUN) of supranational (SNL) and knowledge (REI) of national (NL) nature. An example could be computer games. Finally, the fourth type, Local GIC (LGIC), is characterized by four characteristics of the local nature (NL) but with standardization (GSO) and intellectual property organizations (GIPO) of the global nature. An example could be use of wind power.

6.2 Global Standard Circle

Global Standard Circle (GSC) consists of three main stakeholders:

- Global Standards Body
- Industry
- Societal stakeholders

Global Standards Body (GSB) is a global standardization organization. GSBs are founded by the international community on the highest level, and they develop International Standards. The three biggest and most recognized global standardization organizations are International Organization for Standardization, ISO [3]; International Electrotechnical Commission, IEC [4]; and International Telecommunication Union, ITU [5].

Industry is of a global type, representing all producers, sellers, companies, business and industry associations, foundations, professional bodies, trade associations, etc. Industry has an economic and business interest: to finance the production and to get income out of the production and selling the product that is a result of the standard.

The third main stakeholder is a societal stakeholder in the global standardization process. In general, the societal stakeholders are global consumer organizations and environmental organizations. They ensure relevance of developed goods and services to market expectations regarding environmental protection.

In this chapter, the most important stakeholder, GSB, will be given the highest priority and discussion.

All three of the GSBs, ISO, IEC, and ITU, have existed for more than 50 years (founded in 1947, 1906, and 1865, respectively) with the seat in Geneva, Switzerland. These three organizations have published dozens of thousands of standards, which cover all possible topics. Many of the standards are taken by the change of various non-conformal standards from various economical areas or various countries. On the other side, many of them are created by the joint work of experts in dedicated Technical Committees. The three organizations (ISO, IEC, and ITU) are part of the World Standards Cooperation, WSC, which was founded in 2001. The aim of WSC [6] is strengthening and advancing standardization systems of IEC, ITU, and ISO. WSC focuses to the promotion of the worldwide visibility of the international standardization of the three member organizations. WSC also works on the advancement of the voluntary consensus-based International Standards system. All the member organizations are represented by four current officers. Secretariat and chairmanship rotate annually between the members.

National Standards Bodies, NSBs (i.e., one representative per country), are a part of ISO. IEC has almost the same structure – one representative per country's NSB is the IEC Member. In some cases, the national committee of a certain country in IEC is also a member in ISO. Both organizations, IEC and ISO, are private international organizations not formed by any international contract. Their members can be nongovernmental organizations or governmental agencies, determined by IEC and ISO.

ITU is an organization that was created by a contract between the agency of the United Nations [7] and governments, being the primary members. In addition to countries, other organizations (e.g., nongovernmental or private companies) can hold permanent memberships in ITU.

In addition to these formal organizations, there exist a number of independent global standardization organizations, such as Institute of Electrical and Electronics Engineers, IEEE [8]; Internet Engineering Task Force, IETF [9]; World Wide Web Consortium, W3C [10]; or Universal Postal Union, UPU [11]. These organizations develop and publish standards for various applications. In many cases, these global standardization organizations are not founded with one member per country, but the membership is open to all interested parties, who would like to join the organization, according to the existing rule: either as an organization or as a technical expert.

6.2.1 ISO

International Organization for Standardization (ISO) is an independent, nongovernmental organization with 164 NSOs as member countries representatives. Its General Assembly (GA) is an ultimate ISO authority. The GA is an annual meeting of all members and principal officers. The coordination is made by Central Secretariat in Geneva, Switzerland, and overseen by the Secretary General.

In 1946, in London, the 25 participating delegations agreed on ISO "to facilitate the international coordination and unification of industrial standards." ISO officially started its work in February 1947.

With 164 member countries and 785 Technical Committees and subcommittees, ISO published over 22,631 International Standards in almost all areas of technology and manufacturing.

Figure 6.5 shows distribution of ISO Standards in percentage per technical sector in 2018. Information technology, graphics and photography standards take 21.7% of all ISO Standards, providing the first position of all the ISO sectors.

ISO Statutes define the whole structure of ISO and are written in three languages: English, French, and Russian.

ISO Council consists of up to 20 member bodies with rotation to ensure representation of the member community, ISO Officers, and Chairs of all the Policy Development Committees. ISO Council reports to the GA.

As depicted in Fig. 6.6, four bodies report to the Council:

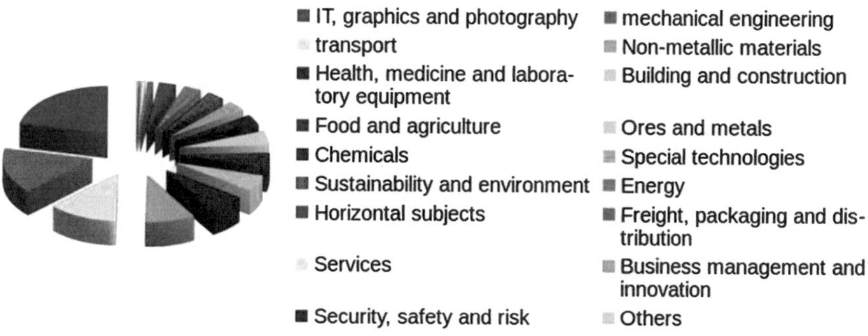

Fig. 6.5 Distribution of ISO Standards per technical sector

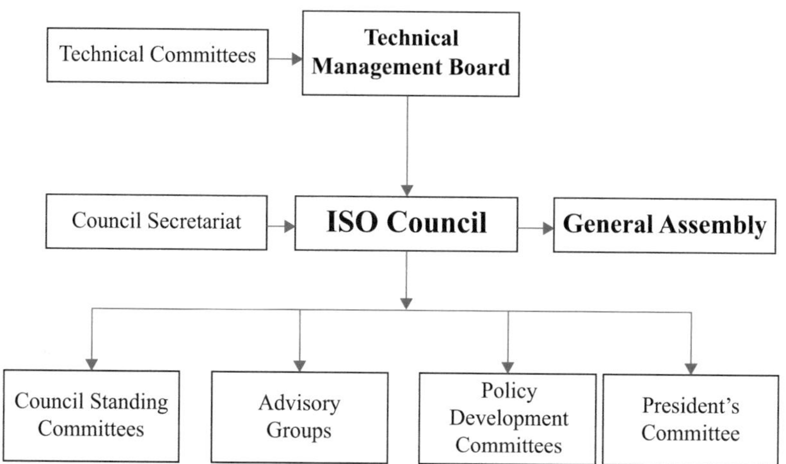

Fig. 6.6 ISO structure

- Council Standing Committees (CSC) are related to finance (CSC/FIN), strategy and policy (CSC/SP), nominations for governance positions (CSC/NOM), and oversight of the ISO governance practices (CSC/OVE).
- Advisory groups are important advisers in the area of ISO's commercial policy (CPAG) and information technology (ITSAG).
- Policy Development Committees: CASCO, providing guidance on conformity assessment; COPOLCO, providing guidance on consumer issues; and DEVCO, providing guidance on matters related to developing countries.
- President's Committee gives advices to the Council related to their area of interest.

Technical Management Board (TMB) manages technical work in ISO. Technical Committees (TCs) develop ISO Standards. TMB is responsible for the work of TCs and for strategic advices related to technical matters. TMB reports to ISO Council.

The list of 12 Technical Committees in the technical sector "Energy" is given in Table 6.2.

Table 6.2 List of 12 ISO Technical Committees in technical sector Energy

Reference	Title	Published standards	Standards under development
ISO/TC 27	Solid mineral fuels	101	18
ISO/TC 28	Petroleum and related products, fuels and lubricants from natural or synthetic sources	257	43
ISO/TC 85	Nuclear energy, nuclear technologies, and radiological protection	219	72
ISO/TC 180	Solar energy	18	3
ISO/TC 193	Natural gas	56	7
ISO/TC 197	Hydrogen technologies	20	7
ISO/TC 238	Solid biofuels	35	12
ISO/TC 255	Biogas	1	3
ISO/TC 263	Coalbed methane (CBM)	2	1
ISO/TC 274	Light and lighting	7	2
ISO/TC 300	Solid Recovered Fuels	0	15
ISO/TC 301	Energy management and energy savings	17	6

Table 6.3 lists 14 TCs in the technical sector "Information technology, graphics and photography."

ISO is a network of National Standards Bodies (NSBs), representing ISO in their country. ISO allows three member categories: full members, correspondent members, and subscriber members.

Full members or member bodies participate and vote in ISO technical and policy meetings. Therefore, they have the influential power to standards development and strategy. Full members sell and adopt ISO International Standards nationally. Correspondent members attend ISO technical and policy meetings as observers. They can sell and adopt ISO International Standards nationally. Subscriber members cannot participate, but rather they keep up to date on ISO work, and they do not sell or adopt ISO International Standards nationally.

As of May 2019, ISO has 164 members whereby 120 are full members, 40 correspondent members, and 4 subscriber members.

As described in Chap. 4, ISO has various types of deliverables. These are ISO Standards, ISO/TS Technical Specifications, ISO/TR Technical Reports, ISO/PAS Publicly Available Specifications, IWA International Workshop Agreements, and ISO Guides.

ISO Standards or ISO International Standards are written to achieve the best possible degree of order in a certain context by providing rules or guidelines activities. Standards exist in many forms. For example, they can be product standards, test methods, codes of practice, or guideline standards. ISO is famous for management systems standards series. As of May 2019, ISO published 22,631 standards.

United Nations (UN) [7] published 17 Sustainable Development Goals (SDGs) and 169 targets that were ratified in September 2015 UN Summit (Table 6.4). SDGs

Table 6.3 List of 14 ISO Technical Committees (TCs) in the technical sector Information technology, graphics and photography

Reference	Title	ISO TC working area	Published standards	Standards under development
ISO/IEC JTC 1	Information technology	Working area	3180	545
ISO/TC 36	Cinematography	Working area	114	4
ISO/TC 42	Photography	Working area	197	16
ISO/TC 46	Information and documentation	Working area	122	20
ISO/TC 130	Graphic technology	Working area	99	36
ISO/TC 154	Processes, data elements and documents in commerce, industry and administration	Working area	25	8
ISO/TC 171	Document management applications	Working area	95	14
ISO/TC 184	Automation systems and integration	Working area	844	46
ISO/TC 211	Geographic information/ Geomatics	Working area	79	21
ISO/TC 213	Dimensional and geometrical product specifications and verification	Working area	147	23
ISO/TC 222	Personal financial planning	Working area	1	0
ISO/PC 288	Educational organizations management systems - Requirements with guidance for use	Working area	1	0
ISO/PC 303	Guidelines on consumer warranties and guarantees	Working area	0	1
ISO/TC 307	Blockchain and distributed ledger technologies	Working area	0	11

aim to play an important role in all sectors (public, nonprofit, for-profit, and voluntary) of the global sustainable development.

Figure 6.7 shows the number of ISO Standards per one SDG, where numbers in Fig. 6.7 correspond to the numbers of the first column in Table 6.4. For example, “1” in Fig. 6.7 means “Goal 1” in Table 6.4.

ISO also develops series of International Standards, such as ISO 9001 on Quality Management, ISO 14001 on Environmental Management, ISO 17000 on Conformity Assessment, ISO/IEC 27001 on Information Security Management, ISO 31000 on Risk Management, ISO 37001 on Anti-bribery Management Systems, ISO 45000 series on Occupational Health and Safety Management Systems, and ISO 50001 on Energy Management. However, external certification bodies, not the ISO or bodies of the ISO, perform the certification.

Table 6.4 UN 17 Sustainable Development Goals

Goal 1	End poverty in all its forms everywhere
Goal 2	End hunger, achieve food security and improved nutrition and promote sustainable agriculture
Goal 3	Ensure healthy lives and promote well-being for all at all ages
Goal 4	Ensure inclusive and equitable quality education and promote lifelong learning opportunities for all
Goal 5	Achieve gender equality and empower all women and girls
Goal 6	Ensure availability and sustainable management of water and sanitation for all
Goal 7	Ensure access to affordable, reliable, sustainable and modern energy for all
Goal 8	Promote sustained, inclusive and sustainable economic growth, full and productive employment and decent work for all
Goal 9	Build resilient infrastructure, promote inclusive and sustainable industrialization and foster innovation
Goal 10	Reduce inequality within and among countries
Goal 11	Make cities and human settlements inclusive, safe, resilient and sustainable
Goal 12	Ensure sustainable consumption and production patterns
Goal 13	Take urgent action to combat climate change and its impacts
Goal 14	Conserve and sustainably use the oceans, seas and marine resources for sustainable development
Goal 15	Protect, restore and promote sustainable use of terrestrial ecosystems, sustainably manage forests, combat desertification, and halt and reverse land degradation and halt biodiversity loss
Goal 16	Promote peaceful and inclusive societies for sustainable development, provide access to justice for all and build effective, accountable and inclusive institutions at all levels
Goal 17	Strengthen the means of implementation and revitalize the global partnership for sustainable development

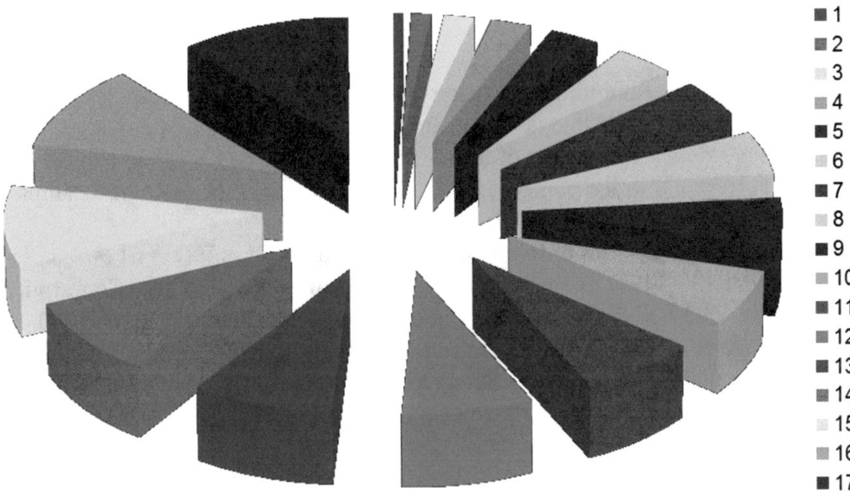

Fig. 6.7 Number of ISO Standards per one SDG

6.2.2 IEC

Before the creation of *International Electrotechnical Commission* (IEC), scientists and engineers struggled with a chaos as they tried to collaborate on emerging discoveries in the electrical industry. Their challenge became painfully clear during the 1904 *International Electrical Congress* held in parallel with the set-up of the Louisiana Purchase Exposition's Palace of Electricity in St. Louis, Missouri, USA (the Exposition was something like a world's fair, although it was organized to celebrate America's 1803 purchase of the Louisiana Territory from France). Each participating country brought exhibits requiring different amounts and types of electrical power. The delegates to the *International Electrical Congress* decided to replace the term "horsepower" with "kilowatt" but still had to reach a common definition of units. Therefore, on September 15, 1904, the delegates concluded: "…steps should be taken to secure the co-operation of the technical societies of the world, by the appointment of a representative Commission to consider the question of the standardization of the nomenclature and ratings of electrical apparatus and machinery." The Commission (IEC) was officially founded in June 1906 in London with the following three study topics: vocabulary, symbols, and electrical machines rating.

Nowadays, the International Electrotechnical Commission is the leading global organization that publishes consensus-based International Standards and manages conformity assessment systems for electric and electronic products, systems, and services, collectively known as electrotechnology. This means that IEC defines safety parameters and processes and parameters of interconnectivity, quality, performance, efficiency, electromagnetic compatibility, and electromagnetic interference. IEC defines an environmental friendliness of a product as well as of recycling together with related symbols and language.

The world's most comprehensive online terminology database on "electrotechnology" is called *Electropedia* or "International Electrotechnical Vocabulary (IEV) online" [12]. Electropedia published a set of publications in the IEC 60050 series. It contains more than 22,000 terminologies in English and French organized by subject area, given in Table 6.5. Equivalent terms are given in various other languages, depending on the area.

As an example, the area 904 "Environmental standardization for electrical and electronic products and systems" is described within four sections. The terms related to the first section, Section 904-01, "General terms relating to the environmental protection and management," can be found in Table 6.6.

Section 904-02 gives terms related to the determination and declaration of substances and materials (Table 6.7).

The next, Section 904-03 presents the terms in the energy efficiency and power consumption (Table 6.8).

Section 904-04 is related to terms in the resource conservation and reuse (Table 6.9).

Table 6.5 Subject areas of Electropedia

Nr	Subject area	Nr	Subject area	Nr	Subject area
101	Mathematics	212	Electrical insulating solids, liquids and gases	444	Elementary relays
102	Mathematics – general concepts and linear algebra	221	Magnetic materials and components	445	Time relays
103	Mathematics – functions	311	Electrical and electronic measurements – general terms relating to measurements	447	Measuring relays
112	Quantities and units	312	Electrical and electronic measurements – general terms relating to electrical measurements	448	Power system protection
113	Physics for electrotechnology	313	Electrical and electronic measurements – types of electrical measuring instruments	461	Electric cables
114	Electrochemistry	314	Electrical and electronic measurements – specific terms according to the type of instrument	466	Overhead lines
121	Electromagnetism	321	Instrument transformers	471	Insulators
131	Circuit theory	351	Control technology	482	Primary and secondary cells and batteries
141	Polyphase systems and circuits	371	Telecontrol	511	Nano-enabled electrotechnical products and systems
151	Electrical and magnetic devices	395	Nuclear instrumentation: physical phenomena, basic concepts, instruments, systems, equipment and detectors	521	Semiconductor devices and integrated circuits
161	Electromagnetic compatibility	411	Rotating machinery	523	Micro-electromechanical devices
171	Digital technology – fundamental concepts	415	Wind turbine generator systems	531	Electronic tubes
192	Dependability	421	Power transformers and reactors	541	Printed circuits
195	Earthing and protection against electric shock	426	Equipment for explosive atmospheres	551	Power electronics
		431	Transduction	561	Piezoelectric devices for frequency control and selection

(continued)

Table 6.5 (continued)

Nr	Subject area	Nr	Subject area	Nr	Subject area
		436	Power capacitors	581	Electromechanical components for electronic equipment
		441	Switchgear, controlgear and fuses	601	Generation, transmission and distribution of electricity - general
		442	Electrical accessories	602	Generation, transmission and distribution of electricity - generation
				603	Generation, transmission and distribution of electricity - power systems planning and management
				605	Generation, transmission and distribution of electricity - substations
				614	Generation, transmission and distribution of electricity - operation
				617	Organization/market of electricity
				651	Live working
				691	Tariffs for electricity
				692	Generation, transmission and distribution of electrical energy – dependability and quality of service of electric power systems

Nr	Subject area	Nr	Subject area	Nr	Subject area
701	Telecommunications, channels and networks	801	Acoustics and electroacoustics	901	Standardization
702	Oscillations, signals and related devices	802	Ultrasonics	902	Conformity assessment
704	Transmission	806	Recording and reproduction of audio and video	903	Risk assessment
705	Radio wave propagation	807	Digital recording of audio and video signals	904	Environmental standardization for electrical and electronic products and systems
712	Antennas	808	Video cameras for non-broadcasting		
713	Radiocommunications: transmitters, receivers, networks and operation	811	Electric traction		

(continued)

Table 6.5 (continued)

Nr	Subject area	Nr	Subject area	Nr	Subject area
714	Switching and signalling in telecommunications	815	Superconductivity		
715	Telecommunication networks, teletraffic and operation	821	Signalling and security apparatus for railways		
716	Integrated services digital network (ISDN) – part 1: General aspects	826	Electrical installations		
721	Telegraphy, facsimile and data communication	841	Industrial electroheat		
722	Telephony	845	Lighting		
723	Broadcasting: Sound, television, data	851	Electric welding		
725	Space radiocommunications	871	Active assisted living (AAL)		
726	Transmission lines and waveguides	881	Radiology and radiological physics		
731	Optical fibre communication	891	Electrobiology		
732	Computer network technology				

Table 6.6 Section 904-01: General terms relating to the environmental protection and management

904-01-01	Environment
904-01-02	Environmental aspect
904-01-03	Environmental impact
904-01-04	Environmental parameter
904-01-05	Process, <in environmental standardization>
904-01-06	Product category
904-01-08	Electronic assembly
904-01-09	Electronic component
904-01-10	Stakeholder
904-01-11	Emission
904-01-12	Design and development
904-01-13	Environmentally conscious design
904-01-14	Environmentally conscious design tool
904-01-15	Environmental management system
904-01-17	End of life
904-01-18	Life cycle assessment
904-01-19	Life cycle stage
904-01-20	Life cycle thinking

Table 6.7 Section 904-02: Terms related to the determination and declaration of substances and materials

904-02-01	Substance
904-02-03	Matrix
904-02-04	Substance group
904-02-05	Declarable substance and declarable substance group
904-02-06	Hazardous mixture
904-02-07	Screening
904-02-08	Reporting threshold level
904-02-09	Performance-based measurement system

Table 6.8 Section 904-03: Terms related to the energy efficiency and power consumption

904-03-01	Power management
904-03-02	Primary function
904-03-03	Secondary function
904-03-04	Tertiary function
904-03-05	Protective function
904-03-06	Reactivation function
904-03-07	Status information function
904-03-08	Network integrity function
904-03-09	Mode
904-03-10	On mode
904-03-11	Partial on mode
904-03-12	Off mode
904-03-13	Operation mode
904-03-14	Idle mode
904-03-15	Disconnected, adj

Table 6.9 Section 904-04: Terms relating to resource conservation and re-use

904-04-01	Disassembly
904-04-02	Disjointment
904-04-03	Energy recovery
904-04-05	Mechanical recycling
904-04-06	Organic recycling
904-04-07	Feedstock recycling
904-04-09	Refurbishing
904-04-10	Remanufacture
904-04-11	Upgrading
904-04-12	Upgradability

The main document of IEC is called *Statutes and Rules of Procedure* [13]. It defines rights and obligations of IEC officers, various IEC management boards, and National Committees members. Technical work is defined in *Directives*, including rules for International Standards.

IEC publications are published in two languages: English and French. Russian Federation National Committee prepares Russian language editions.

IEC documents fall into two categories: *normative publications* that are related to technical description of characteristics of a product, system, service, or object and *informative publications* that are informative in the form of implementation procedures or guidelines.

IEC has various types of deliverables. These are IEC International Standards, IEC/TS Technical Specifications, IEC/TR Technical Reports, IEC/PAS Publicly Available Specifications, IEC Amendments, IEC Technical Corrigenda, IEC Interpretation Sheets, and IEC Guides. As explained for all the IEC documents in the Chap. 4, IEC publishes them all in the field of electrotechnology.

The IEC International Standards (ISs) are of an especial value for all devices using, producing, or storing electricity or containing any kind of electronics. As given in the Chap. 4, in Sect. 4.2.8b, IEC fields of activity are quite wide, due to the exponential development of electrical engineering in general. The covered areas are, for example, electronics, electrical equipment and batteries, household devices, electrical vehicles, electrical generators and power lines as well as nanotechnology.

In Table 6.10 are written published IEC Guides, covering topics from "Risk management - Vocabulary" (ISO/IEC GUIDE 98-3-SP1:2008/COR1:2009 [53]) to the ISO/IEC GUIDE 98-3:2008 [51], which covers topic of uncertainty of measurement as a guide to the expression of uncertainty in measurement.

Some of the guides are designated by SMB to be normative. They are as follows:

- *IEC Guide 104 Ed. 6.0 (2010-08)* [60] "The preparation of safety publications and the use of basic safety publications and group safety publications"
- *IEC Guide 107 Ed. 6.0 (2014-07)* [61] "Electromagnetic compatibility - Guide to the drafting of electromagnetic compatibility publications"
- *IEC Guide 108 Ed. 2.0 (2006-08)* [62] "Guidelines for ensuring the coherency of IEC publications - Application of horizontal standards"

The standards are results of *IEC Technical Committees* and Subcommittees (TCs/SCs). As of February 2020, there are 208 TC/SC, 680 Working Groups (WGs), 205 Project Teams (PTs), and 646 Maintenance Teams (MT). The list is given in Table 6.11. The vast number of topics is covered: from Terminology (TC 1) to Audio, video and multimedia systems and equipment (TC 100). The number of publications per TC looks sometimes incredible (594) in the ISO/IEC JTC 1/SC 29, dealing with the coding of audio, picture, multimedia, and hypermedia information.

An incredible number of 21,129 IEC technical experts are involved in the IEC TCs. The technical experts have different backgrounds, coming from governments, all areas of the industry, academia, test and research laboratories, and consumer groups. Their common goal is to represent the national electrotechnical needs of IEC Member countries (Table 6.12) and IEC Affiliates (Table 6.13) in order to develop globally relevant, voluntary, neutral, independent, and consensus-based IEC International Standards. It is important that participants have both certain and broad competencies, and devotion. There is a lot of voluntary and enthusiastic work of many professionals in standards development.

Table 6.10 Published IEC Guides

Committee	Reference	Title
SMB	IEC GUIDE 103:1980 [14]	Guide on dimensional co-ordination
ACOS	IEC GUIDE 104:2010 [15]	The preparation of safety publications and the use of basic safety publications and group safety publications
ACEC	IEC GUIDE 107:2014 [16]	Electromagnetic compatibility - Guide to the drafting of electromagnetic compatibility publications
SMB	IEC GUIDE 108:2006 [17]	Guidelines for ensuring the coherency of IEC publications - Application of horizontal standards
ACEA	IEC GUIDE 109:2012 [18]	Environmental aspects - Inclusion in electrotechnical product standards
ACOS	IEC GUIDE 110:2014 [19]	Home control systems - Guidelines relating to safety
ACTAD	IEC GUIDE 111:2004 [20]	Electrical high-voltage equipment in high-voltage substations - Common recommendations for product standards
ACOS	IEC GUIDE 112:2017 [21]	Guide on the safety of multimedia equipment
IECEE-CTL	IEC GUIDE 115:2007 [22]	Application of uncertainty of measurement to conformity assessment activities in the electrotechnical sector
ACOS	IEC GUIDE 116:2018 [23]	Guidelines for safety related risk assessment and risk reduction for low voltage equipment
ACOS	IEC GUIDE 117:2010 [24]	Electrotechnical equipment - Temperatures of touchable hot surfaces
ACEE	IEC GUIDE 118:2017 [25]	Inclusion of energy efficiency aspects in electrotechnical publications
ACEE	IEC GUIDE 119:2017 [26]	Preparation of energy efficiency publications and the use of basic energy efficiency publications and group energy efficiency publications
ACSEC	IEC GUIDE 120:2018 [27]	Security aspects - Guidelines for their inclusion in publications
ISO/TMB	ISO/IEC GUIDE 2:2004 [28]	Standardization and related activities – General vocabulary
ISO/ COPOLCO	ISO/IEC GUIDE 14:2018 [29]	Products and related services - Information for consumers
ISO/TMB	ISO/IEC GUIDE 17:2016 [30]	Guide for writing standards taking into account the needs of micro, small and medium-sized enterprises
ISO/TMB	ISO/IEC GUIDE 21-1:2005 [31]	Regional or national adoption of International Standards and other International Deliverables – Part 1: Adoption of International Standards
ISO/TMB	ISO/IEC GUIDE 21-2:2005 [32]	Regional or national adoption of International Standards and other International Deliverables – Part 2: Adoption of International Deliverables other than International Standards
ISO/ CASCO	ISO/IEC GUIDE 23:1982 [33]	Methods of indicating conformity with standards for third-party certification systems

(continued)

Table 6.10 (continued)

Committee	Reference	Title
ISO/ COPOLCO	ISO/IEC GUIDE 37:2012 [34]	Instructions for use of products by consumers
ISO/ COPOLCO	ISO/IEC GUIDE 41:2018 [35]	Packaging - Recommendations for addressing consumer needs
ISO/ COPOLCO	ISO/IEC GUIDE 46:2017 [36]	Comparative testing of consumer products and related services – General principles
ISO/ COPOLCO	ISO/IEC GUIDE 50:2014 [37]	Safety aspects - Guidelines for child safety in standards and other specifications
ISO/ COPOLCO	ISO/IEC GUIDE 51:2014 [38]	Safety aspects – Guidelines for their inclusion in standards
ISO/TMB	ISO/IEC GUIDE 59:1994 [39]	Code of good practice for standardization
ISO/ CASCO	ISO/IEC GUIDE 60:2004 [40]	Conformity assessment – Code of good practice
ISO/TC 210	ISO/IEC GUIDE 63:2012 [41]	Guide to the development and inclusion of safety aspects in International Standards for medical devices
ISO/ CASCO	ISO/IEC GUIDE 68:2002 [42]	Arrangements for the recognition and acceptance of conformity assessment results
ISO/TMB	ISO/IEC GUIDE 71:2014 [43]	Guide for addressing accessibility in standards
ISO/ COPOLCO	ISO/IEC GUIDE 74:2004 [44]	Graphical symbols – Technical guidelines for the consideration of consumers' needs
SMB	ISO/IEC GUIDE 75:2006 [45]	Strategic principles for future IEC and ISO standardization in industrial automation
ISO/ COPOLCO	ISO/IEC GUIDE 76:2008 [46]	Development of service standards – Recommendations for addressing consumer issues
ISO/TMB	ISO/IEC GUIDE 77-1:2008 [47]	Guide for specification of product properties and classes – Part 1: Fundamental benefits
ISO/TMB	ISO/IEC GUIDE 77-2:2008 [48]	Guide for specification of product properties and classes – Part 2: Technical principles and guidance
ISO/TMB	ISO/IEC GUIDE 77-3:2008 [49]	Guide for specification of product properties and classes – Part 3: Experience gained
ISO/TMB	ISO/IEC GUIDE 98-1:2009 [50]	Uncertainty of measurement - Part 1: Introduction to the expression of uncertainty in measurement
ISO/TMB	ISO/IEC GUIDE 98-3:2008 [51]	Uncertainty of measurement – Part 3: Guide to the expression of uncertainty in measurement (GUM:1995)
ISO/TMB	ISO/IEC GUIDE 98-3-SP1:2008 [52]	Supplement 1 - Uncertainty of measurement - Part 3: Guide to the expression of uncertainty in measurement (GUM:1995) - Propagation of distributions using a Monte Carlo method
ISO/TMB	ISO/IEC GUIDE 98-3-SP1:2008/ COR1:2009 [53]	Corrigendum 1 - Supplement 1 - Uncertainty of measurement - Part 3: Guide to the expression of uncertainty in measurement (GUM:1995) - Supplement 1: Propagation of distributions using a Monte Carlo method

(continued)

Table 6.10 (continued)

Committee	Reference	Title
ISO/TMB	ISO/IEC GUIDE 98-3-SP2:2011 [54]	Supplement 2 - Uncertainty of measurement – Part 3: Guide to the expression of uncertainty in measurement (GUM:1995) - Extension to any number of output quantities
ISO/TMB	ISO/IEC GUIDE 98-4:2012 [55]	Uncertainty of measurement – Part 4: Role of measurement uncertainty in conformity assessment
ISO/TMB	ISO/IEC GUIDE 99:2007 [56]	International vocabulary of metrology – Basic and general concepts and associated terms (VIM)
ISO/TC 207	ISO GUIDE 64:2008 [57]	Guide for addressing environmental issues in product standards
ISO/TMB	ISO GUIDE 72:2001 [58]	Guidelines for the justification and development of management system standards
ISO/TC 262	ISO GUIDE 73:2009 [59]	Risk management - Vocabulary

Table 6.11 List of IEC TC and SC

Committee	Description	Publications
TC 1	Terminology	194
TC 2	Rotating machinery	71
TC 3	Information structures and elements, identification and marking principles, documentation and graphical symbols	40
SC 3C	Graphical symbols for use on equipment	13
SC 3D	Product properties and classes and their identification	8
TC 4	Hydraulic turbines	29
TC 5	Steam turbines	6
TC 7	Overhead electrical conductors	21
TC 8	System aspects of electrical energy supply	11
SC 8A	Grid Integration of Renewable Energy Generation	0
SC 8B	Decentralized Electrical Energy Systems	2
TC 9	Electrical equipment and systems for railways	144
TC 10	Fluids for electrotechnical applications	61
TC 11	Overhead lines	11
TC 13	Electrical energy measurement and control	86
TC 14	Power transformers	42
TC 15	Solid electrical insulating materials	240
TC 17	High-voltage switchgear and controlgear	4
SC 17A	Switching devices	43
SC 17C	Assemblies	22
TC 18	Electrical installations of ships and of mobile and fixed offshore units	48
SC 18A	Electric cables for ships and mobile and fixed offshore units	10
TC 20	Electric cables	243

(continued)

Table 6.11 (continued)

Committee	Description	Publications
TC 21	Secondary cells and batteries	37
SC 21A	Secondary cells and batteries containing alkaline or other non-acid electrolytes	22
TC 22	Power electronic systems and equipment	11
SC 22E	Stabilized power supplies	10
SC 22F	Power electronics for electrical transmission and distribution systems	61
SC 22G	Adjustable speed electric drive systems incorporating semiconductor power converters	25
SC 22H	Uninterruptible power systems (UPS)	12
TC 23	Electrical accessories	25
SC 23A	Cable management systems	29
SC 23B	Plugs, socket-outlets and switches	58
SC 23E	Circuit-breakers and similar equipment for household use	61
SC 23G	Appliance couplers	15
SC 23H	Plugs, Socket-outlets and Couplers for industrial and similar applications, and for Electric Vehicles	23
SC 23J	Switches for appliances	12
SC 23K	Electrical energy efficiency products	0
TC 25	Quantities and units	27
TC 26	Electric welding	25
TC 27	Industrial electroheating and electromagnetic processing	29
TC 29	Electroacoustics	73
TC 31	Equipment for explosive atmospheres	53
SC 31G	Intrinsically-safe apparatus	9
SC 31J	Classification of hazardous areas and installation requirements	12
SC 31M	Non-electrical equipment and protective systems for explosive atmospheres	8
TC 32	Fuses	3
SC 32A	High-voltage fuses	7
SC 32B	Low-voltage fuses	21
SC 32C	Miniature fuses	19
TC 33	Power capacitors and their applications	40
TC 34	Lamps and related equipment	49
SC 34A	Lamps	164
SC 34B	Lamp caps and holders	216
SC 34C	Auxiliaries for lamps	72
SC 34D	Luminaires	59
TC 35	Primary cells and batteries	12
TC 36	Insulators	51
SC 36A	Insulated bushings	7
TC 37	Surge arresters	8

(continued)

Table 6.11 (continued)

Committee	Description	Publications
SC 37A	Low-voltage surge protective devices	11
SC 37B	Components for low-voltage surge protection	8
TC 38	Instrument transformers	21
TC 40	Capacitors and resistors for electronic equipment	163
TC 42	High-voltage and high-current test techniques	14
TC 44	Safety of machinery - Electrotechnical aspects	34
TC 45	Nuclear instrumentation	28
SC 45A	Instrumentation, control and electrical power systems of nuclear facilities	97
SC 45B	Radiation protection instrumentation	56
TC 46	Cables, wires, waveguides, RF connectors, RF and microwave passive components and accessories	58
SC 46A	Coaxial cables	78
SC 46C	Wires and symmetric cables	74
SC 46F	RF and microwave passive components	110
TC 47	Semiconductor devices	116
SC 47A	Integrated circuits	84
SC 47D	Semiconductor devices packaging	57
SC 47E	Discrete semiconductor devices	44
SC 47F	Micro-electromechanical systems	34
TC 48	Electrical connectors and mechanical structures for electrical and electronic equipment	1
SC 48B	Electrical connectors	250
SC 48D	Mechanical structures for electrical and electronic equipment	38
TC 49	Piezoelectric, dielectric and electrostatic devices and associated materials for frequency control, selection and detection	92
TC 51	Magnetic components, ferrite and magnetic powder materials	72
TC 55	Winding wires	132
TC 56	Dependability	60
TC 57	Power systems management and associated information exchange	188
TC 59	Performance of household and similar electrical appliances	13
SC 59A	Electric dishwashers	2
SC 59C	Electrical heating appliances for household and similar purposes	14
SC 59D	Performance of household and similar electrical laundry appliances	13
SC 59F	Surface cleaning appliances	9
SC 59K	Performance of household and similar electrical cooking appliances	14
SC 59L	Small household appliances	32
SC 59M	Performance of electrical household and similar cooling and freezing appliances	7
TC 61	Safety of household and similar electrical appliances	292

(continued)

Table 6.11 (continued)

Committee	Description	Publications
SC 61B	Safety of microwave appliances for household and commercial use	7
SC 61C	Safety of refrigeration appliances for household and commercial use	19
SC 61D	Appliances for air-conditioning for household and similar purposes	4
SC 61H	Safety of electrically-operated farm appliances	14
SC 61J	Electrical motor-operated cleaning appliances for commercial use	13
TC 62	Electrical equipment in medical practice	1
SC 62A	Common aspects of electrical equipment used in medical practice	63
SC 62B	Diagnostic imaging equipment	78
SC 62C	Equipment for radiotherapy, nuclear medicine and radiation dosimetry	40
SC 62D	Electromedical equipment	108
TC 64	Electrical installations and protection against electric shock	75
TC 65	Industrial-process measurement, control and automation	33
SC 65A	System aspects	52
SC 65B	Measurement and control devices	105
SC 65C	Industrial networks	157
SC 65E	Devices and integration in enterprise systems	88
TC 66	Safety of measuring, control and laboratory equipment	38
TC 68	Magnetic alloys and steels	61
TC 69	Electric road vehicles and electric industrial trucks	21
TC 70	Degrees of protection provided by enclosures	14
TC 72	Automatic electrical controls	29
TC 73	Short-circuit currents	14
TC 76	Optical radiation safety and laser equipment	35
TC 77	Electromagnetic compatibility	18
SC 77A	EMC – low frequency phenomena	85
SC 77B	High frequency phenomena	28
SC 77C	High power transient phenomena	24
TC 78	Live working	66
TC 79	Alarm and electronic security systems	53
TC 80	Maritime navigation and radiocommunication equipment and systems	70
TC 81	Lightning protection	17
TC 82	Solar photovoltaic energy systems	119
TC 85	Measuring equipment for electrical and electromagnetic quantities	83
TC 86	Fibre optics	27
SC 86A	Fibres and cables	105
SC 86B	Fibre optic interconnecting devices and passive components	252
SC 86C	Fibre optic systems and active devices	128
TC 87	Ultrasonics	55

(continued)

Table 6.11 (continued)

Committee	Description	Publications
TC 88	Wind energy generation systems	30
TC 89	Fire hazard testing	49
TC 90	Superconductivity	24
TC 91	Electronics assembly technology	186
TC 94	All-or-nothing electrical relays	14
TC 95	Measuring relays and protection equipment	15
TC 96	Transformers, reactors, power supply units, and combinations thereof	23
TC 97	Electrical installations for lighting and beaconing of aerodromes	7
TC 99	Insulation co-ordination and system engineering of high voltage electrical power installations above 1,0 kV AC and 1,5 kV DC	13
TC 100	Audio, video and multimedia systems and equipment	97
TA 1	Terminals for audio, video and data services and contents	38
TA 2	Colour measurement and management	19
TA 4	Digital system interfaces and protocols	33
TA 5	Cable networks for television signals, sound signals and interactive services	26
TA 6	Storage media, storage data structures, storage systems and equipment	140
TA 10	Multimedia e-publishing and e-book technologies	11
TA 15	Wireless power transfer	5
TA 16	Active Assisted Living (AAL), wearable electronic devices and technologies, accessibility and user interfaces	7
TA 17	Multimedia systems and equipment for cars	3
TA 18	Multimedia home systems and applications for end-user networks	46
TA 19	Environmental and energy aspects for multimedia systems and equipment	13
TA 20	Analogue and digital audio	79
TC 101	Electrostatics	35
TC 103	Transmitting equipment for radiocommunication	37
TC 104	Environmental conditions, classification and methods of test	129
TC 105	Fuel cell technologies	22
TC 106	Methods for the assessment of electric, magnetic and electromagnetic fields associated with human exposure	30
TC 107	Process management for avionics	26
TC 108	Safety of electronic equipment within the field of audio/video, information technology and communication technology	32
TC 109	Insulation co-ordination for low-voltage equipment	12
TC 110	Electronic displays	159
TC 111	Environmental standardization for electrical and electronic products and systems	25
TC 112	Evaluation and qualification of electrical insulating materials and systems	75

(continued)

Table 6.11 (continued)

Committee	Description	Publications
TC 113	Nanotechnology for electrotechnical products and systems	37
TC 114	Marine energy – wave, tidal and other water current converters	13
TC 115	High Voltage Direct Current (HVDC) transmission for DC voltages above 100 kV	9
TC 116	Safety of motor-operated electric tools	125
TC 117	Solar thermal electric plants	4
PC 118	Smart grid user interface	4
TC 119	Printed electronics	20
TC 120	Electrical energy storage (EES) systems	6
TC 121	Switchgear and controlgear and their assemblies for low voltage	0
SC 121A	Low-voltage switchgear and controlgear	78
SC 121B	Low-voltage switchgear and controlgear assemblies	14
TC 122	UHV AC transmission systems	4
TC 123	Management of network assets in power systems	0
TC 124	Wearable electronic devices and technologies	0
TC 125	Personal e-Transporters (PeTs)	0
PC 126	Binary power generation systems	0
PC 127	Low-voltage auxiliary power systems for electric power plants and substations	0
CISPR	International special committee on radio interference	0
CIS/A	Radio-interference measurements and statistical methods	45
CIS/B	Interference relating to industrial, scientific and medical radio-frequency apparatus, to other (heavy) industrial equipment, to overhead power lines, to high voltage equipment and to electric traction	13
CIS/D	Electromagnetic disturbances related to electric/electronic equipment on vehicles and internal combustion engine powered devices	5
CIS/F	Interference relating to household appliances tools, lighting equipment and similar apparatus	9
CIS/H	Limits for the protection of radio services	11
CIS/I	Electromagnetic compatibility of information technology equipment, multimedia equipment and receivers	12
CIS/S	Steering Committee	0
SyC AAL	Active Assisted Living	0
SyC LVDC	Low voltage direct current and low voltage direct current for electricity access	0
SyC SM	Smart manufacturing	0
SyC Smart Cities	Electrotechnical aspects of smart cities	0
SyC Smart Energy	Smart energy	8
ISO/IEC JTC 1	Information technology	538

(continued)

Table 6.11 (continued)

Committee	Description	Publications
ISO/IEC JTC 1/SC 2	Coded character sets	51
ISO/IEC JTC 1/SC 6	Telecommunications and information exchange between systems	391
ISO/IEC JTC 1/SC 7	Software and systems engineering	186
ISO/IEC JTC 1/SC 17	Cards and security devices for personal identification	109
ISO/IEC JTC 1/SC 22	Programming languages, their environments and system software interfaces	104
ISO/IEC JTC 1/SC 23	Digitally recorded media for information interchange and storage	139
ISO/IEC JTC 1/SC 24	Computer graphics, image processing and environmental data representation	81
ISO/IEC JTC 1/SC 25	Interconnection of information technology equipment	216
ISO/IEC JTC 1/SC 27	IT security techniques	185
ISO/IEC JTC 1/SC 28	Office equipment	48
ISO/IEC JTC 1/SC 29	Coding of audio, picture, multimedia and hypermedia information	594
ISO/IEC JTC 1/SC 31	Automatic identification and data capture techniques	118
ISO/IEC JTC 1/SC 32	Data management and interchange	76
ISO/IEC JTC 1/SC 34	Document description and processing languages	79
ISO/IEC JTC 1/SC 35	User interfaces	74
ISO/IEC JTC 1/SC 36	Information technology for learning, education and training	43
ISO/IEC JTC 1/SC 37	Biometrics	116
ISO/IEC JTC 1/SC 38	Cloud computing and distributed platforms	15
ISO/IEC JTC 1/SC 39	Sustainability for and by information technology	19
ISO/IEC JTC 1/SC 40	IT service management and IT governance	21
ISO/IEC JTC 1/SC 41	Internet of things and related technologies	20
ISO/IEC JTC 1/SC 42	Artificial intelligence	1

Table 6.12 IEC Membership

Country	Code	IEC Membership	Website
Albania	AL	Associate Member	http://www.dps.gov.al
Algeria	DZ	Full Member	http://www.ianor.dz
Argentina	AR	Full Member	http://www.aea.org.ar
Australia	AU	Full Member	http://www.standards.org.au
Austria	AT	Full Member	http://www.ove.at
Bahrain	BH	Associate Member	http://www.moic.gov.bh
Bangladesh	BD	Associate Member	http://www.bsti.gov.bd
Belarus	BY	Full Member	http://www.gosstandart.gov.by
Belgium	BE	Full Member	http://www.ceb-bec.be
Bosnia and Herzegovina	BA	Associate Member	http://www.bas.gov.ba/button_82.html
Brazil	BR	Full Member	http://www.cobei.org.br
Bulgaria	BG	Full Member	http://www.bds-bg.org
Canada	CA	Full Member	http://www.scc.ca
Chile	CL	Full Member	http://www.cornelec.cl
China	CN	Full Member	http://www.sac.gov.cn
Colombia	CO	Full Member	http://www.icontec.org
Croatia	HR	Full Member	http://www.hzn.hr
Cuba	CU	Associate Member	http://www.nc.cubaindustria.cu/
Cyprus	CY	Associate Member	http://www.cys.org.cy
Czech Republic	CZ	Full Member	http://www.unmz.cz/
Côte d'Ivoire	CI	Associate Member	http://www.codinorm.ci/
Democratic People's Republic of Korea	KP	Associate Member	
Denmark	DK	Full Member	http://www.ds.dk
Egypt	EG	Full Member	http://www.moee.gov.eg/
Estonia	EE	Associate Member	http://www.evs.ee
Finland	FI	Full Member	http://www.sesko.fi
France	FR	Full Member	http://www.afnor.org
Georgia	GE	Associate Member	http://www.geostm.ge
Germany	DE	Full Member	http://www.dke.de
Ghana	GH	Associate Member	https://gsa.gov.gh
Greece	GR	Full Member	http://www.elot.gr
Hungary	HU	Full Member	http://www.mszt.hu

(continued)

Table 6.12 (continued)

Country	Code	IEC Membership	Website
Iceland	IS	Associate Member	http://www.stadlar.is
India	IN	Full Member	http://www.bis.org.in
Indonesia	ID	Full Member	http://bsn.go.id
Iran	IR	Full Member	http://www.inec.ir
Iraq	IQ	Full Member	http://cosqc.gov.iq/en/home.aspx
Ireland	IE	Full Member	http://www.nsai.ie
Israel	IL	Full Member	http://www.sii.org.il
Italy	IT	Full Member	http://www.ceinorme.it
Japan	JP	Full Member	http://www.jisc.go.jp/
Jordan	JO	Associate Member	http://www.jsmo.gov.jo
Kazakhstan	KZ	Associate Member	http://www.memst.kz
Kenya	KE	Associate Member	http://www.kebs.org
Korea, Republic of	KR	Full Member	http://www.kats.go.kr/
Kuwait	KW	Full Member	http://www.pai.gov.kw
Latvia	LV	Associate Member	http://www.lvs.lv
Lithuania	LT	Associate Member	http://www.lsd.lt
Luxembourg	LU	Full Member	http://www.portail-qualite.lu/
Malaysia	MY	Full Member	http://www.jsm.gov.my
Malta	MT	Associate Member	http://www.mccaa.org.mt/
Mexico	MX	Full Member	http://www.economia.gob.mx/?P=85
Moldova	MD	Associate Member	http://www.standard.md
Montenegro	ME	Associate Member	http://www.isme.me/
Morocco	MA	Associate Member	http://www.imanor.ma/
Netherlands	NL	Full Member	http://www.nen.nl
New Zealand	NZ	Full Member	http://www.standards.govt.nz/
Nigeria	NG	Full Member	http://www.son.gov.ng
North Macedonia	MK	Associate Member	http://www.isrm.gov.mk
Norway	NO	Full Member	http://www.nek.no
Oman	OM	Full Member	
Pakistan	PK	Full Member	http://www.psqca.com.pk
Peru	PE	Full Member	http://www.inacal.gob.pe
Philippines, Rep. of the	PH	Full Member	http://www.bps.dti.gov.ph

(continued)

Table 6.12 (continued)

Country	Code	IEC Membership	Website
Poland	PL	Full Member	http://www.pkn.pl
Portugal	PT	Full Member	http://www.ipq.pt
Qatar	QA	Full Member	
Romania	RO	Full Member	http://www.asro.ro
Russian Federation	RU	Full Member	http://www.gost.ru
Saudi Arabia	SA	Full Member	http://www.saso.gov.sa
Serbia	RS	Full Member	http://www.iss.rs
Singapore	SG	Full Member	http://www.enterprisesg.gov.sg
Slovakia	SK	Full Member	http://www.unms.sk
Slovenia	SI	Full Member	http://www.sist.si
South Africa	ZA	Full Member	www.sabs.co.za/sanc
Spain	ES	Full Member	http://www.une.org
Sri Lanka	LK	Associate Member	http://www.nsf.ac.lk
Sweden	SE	Full Member	http://www.elstandard.se
Switzerland	CH	Full Member	http://www.electrosuisse.ch
Thailand	TH	Full Member	http://www.tisi.go.th
Tunisia	TN	Associate Member	http://www.innorpi.tn
Turkey	TR	Full Member	http://www.tse.org.tr
Uganda	UG	Associate Member	http://unbs.go.ug
Ukraine	UA	Full Member	http://www.uas.org.ua
United Arab Emirates	AE	Full Member	http://www.esma.gov.ae
United Kingdom	GB	Full Member	http://www.bsigroup.com
United States of America	US	Full Member	http://www.ansi.org
Vietnam	VN	Associate Member	http://www.tcvn.gov.vn

Table 6.13 IEC Affiliate Countries

Country	Code	Website
Afghanistan	AF	http://ansa.gov.af/
Angola	AO	www.ianorq.gov.ao
Antigua and Barbuda	AG	http://abbs.gov.ag/
Armenia	AM	http://www.sarm.am/en
Azerbaijan	AZ	http://azstand.gov.az/en
Bahamas	BS	http://www.bbsq.bs/en/
Barbados	BB	http://www.bnsi.bb/
Belize	BZ	http://www.bbs.gov.bz
Benin	BJ	anmbenin.com/
Bhutan	BT	http://www.bsb.gov.bt/

(continued)

Table 6.13 (continued)

Country	Code	Website
Bolivia	BO	http://www.ibnorca.org/
Botswana	BW	http://www.bobstandards.bw
Brunei Darussalam	BN	http://www.mod.gov.bn/
Burkina Faso	BF	http://www.abnorm.bf/
Burundi	BI	http://www.bbn-burundi.org/
Cabo Verde	CV	http://www.igqpi.cv
Cambodia	KH	http://www.isc.gov.kh
Cameroon	CM	http://www.anorcameroun.info
Central African Republic	CF	sites.google.com/a/minco-rca.org/www/
Chad	TD	http://www.commerce.gouv.td
Comoros	KM	
Congo	CG	
Costa Rica	CR	http://www.inteco.org
Democratic Republic of the Congo	CD	http://occ.cd/
Djibouti, Republic of	DJ	
Dominica	DM	
Dominican Republic	DO	http://www.indocal.gob.do
Ecuador	EC	http://www.normalizacion.gob.ec
El Salvador	SV	http://www.osa.gob.sv
Eritrea	ER	
Eswatini, Kingdom of	SZ	http://www.swasa.co.sz
Ethiopia	ET	http://www.ethiostandards.org
Fiji	FJ	http://www.mit.gov.fj/
Gabon	GA	
Gambia	GM	
Grenada	GD	http://www.gdbs.gd/
Guatemala	GT	http://www.mineco.gob.gt
Guinea	GN	
Guinea-Bissau	GW	
Guyana	GY	http://www.gnbsgy.org
Haiti	HT	
Honduras	HN	http://ohn.hondurascalidad.org
Jamaica	JM	http://www.bsj.org.jm
Kyrgyzstan	KG	http://www.nism.gov.kg/en
Lao People's Democratic Republic	LA	http://www.most.gov.la/index.php/en
Lebanon	LB	http://www.oea.org.lb
Lesotho	LS	
Liberia	LR	http://www.moci.gov.lr
Madagascar	MG	http://www.bnm.mg
Malawi	MW	http://www.mbsmw.org
Mali	ML	
Mauritania	MR	http://www.dnpq.mr

(continued)

Table 6.13 (continued)

Country	Code	Website
Mauritius	MU	http://msb.intnet.mu
Mongolia	MN	http://masm.gov.mn
Mozambique	MZ	http://www.innoq.gov.mz
Myanmar	MM	http://www.most.gov.mm/most2eng/
Namibia	NA	http://www.nsi.com.na
Nepal	NP	http://www.nbsm.gov.np/
Niger	NE	
Palestine	PS	http://www.psi.pna.ps
Panama	PA	http://www.mici.gob.pa
Papua New Guinea	PG	http://www.nisit.gov.pg
Paraguay	PY	http://www.intn.gov.py
Rwanda	RW	http://www.rsb.gov.rw
Saint Kitts and Nevis	KN	
Saint Lucia	LC	http://www.slbs.org.lc
Saint Vincent and the Grenadines	VC	http://www.svgbs.gov.vc
São Tomé and Príncipe	ST	
Senegal	SN	http://www.asn.sn
Seychelles	SC	http://www.seychelles.net/sbsorg
Sierra Leone	SL	http://slstandards.org/
South Sudan, The Republic of	SS	
Sudan	SD	http://www.ssmo.gov.sd/home.php
Suriname	SR	http://www.ssb.sr
Syrian Arab Republic	SY	http://www.sasmo.org.sy/en/
Tanzania	TZ	http://www.tbs.go.tz/
Togo	TG	
Trinidad and Tobago	TT	http://www.ttbs.org.tt
Turkmenistan	TM	
Uruguay	UY	http://www.unit.org.uy
Uzbekistan	UZ	http://www.standart.uz
Yemen	YE	
Zambia	ZM	http://www.zabs.org.zm/
Zimbabwe	ZW	http://www.saz.org.zw

Full membership means that the NC has all the rights in terms of free access to all technical and management activities and functions, including voting rights in the IEC Council. Associate membership means that the NC has limited voting rights in the technical work, no eligibility to managerial functions, but a full access to all working documents. As of February 2020, IEC has 62 full members and 26 associate members, which gives a total of 88 member countries.

European Union members of IEC are *National Committees* (NCs). IEC recognizes only one NC per country. Individuals participate in the IEC's work through the NCs. Each NC agrees to an open-access and balanced full representation of private and public electrotechnical interest of the represented country in terms of

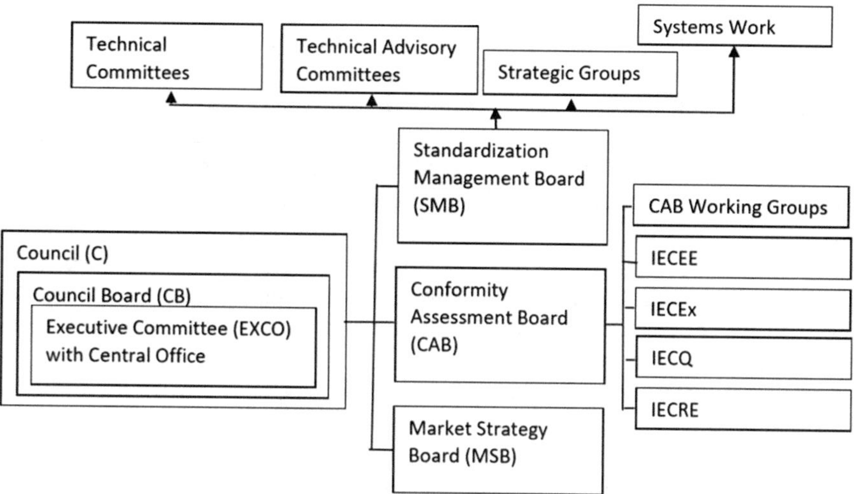

Fig. 6.8 IEC structure

standardization and conformity assessment in order to enable all stakeholders a real influence on the IEC work.

IEC Affiliate Countries enabled enlargement of the IEC Membership. *Affiliate members* develop countries that do not bear any financial burden due to the IEC Membership, but they enjoy full use of IEC electronic environment. The *World Trade Organization* (WTO) [63] complements IEC to enhance participation from developing countries in the standardization process, which results in economic development of the countries by removing technical barriers to trade.

IEC NCs, as members of the TCs, can participate in the work of any TC as *participating members* (P-Members) or *observer members* (O-Members). Participating members are active contributors in the meetings. They have the obligation to vote at all stages of standards development. Observer members follow the work by receiving committee documents, and they have the right to attend meetings and submit comments.

Figure 6.8 shows the IEC structure. The *IEC Council* (C) is the supreme governing legislative body of the IEC that meets at least once a year at the IEC General Meeting (since 1904). It is responsible for an application approval of the IEC Membership and proposed amendments to IEC Statutes and Rules of Procedure. Presidents of all IEC Full Member National Committees, current IEC Officers, all Past Presidents, and Council Board members are IEC Council members. Only presidents of full member NCs have voting rights, meaning that one country has only one vote. Observers are usually presidents of associate member NCs, since they have no voting rights. The IEC Council sets the IEC long-term strategic and financial objectives, and it elects IEC officers. The Council delegates the management of IEC daily operational work to IEC Council Board (CB). The Standardization Management Board (SMB) performs specific management in the area of IEC standards work. The Conformity Assessment Board (CAB) performs specific tasks in the area of conformity assessment, while the Market Strategy Board (MSB) covers the third area of market strategy. The IEC Council elects members of CB, SMB, and CAB.

The Council Board (CB) recommends IEC policy documents to the Council and implements Council policy. The CB is also responsible for decisions on the operational level, except for finances. CB receives reports from three sides: from SMB, CAB, and MSB. If needed, it can set up ad hoc working groups for a specific topic. The CB is a high-level decision-making body that sends all the documents of interest to the IEC Council. Members of CB are Chair (IEC President) and 15 members elected by the Council. The IEC officers are without vote, acting as ex officio members. The CB members act like individuals without representation of their NC or any industry or association. Their mission on CB is only to work for the IEC community. The CB meets at least twice a year.

The Standardization Management Board (SMB) is a body that makes decisions and reports to CB in the area of IEC standards work. SMB consists of IEC Vice-President, being SMB Chair, 15 members elected by Council, and 7 permanent members appointed from NCs paying the highest percentage of annual dues combined with the highest number of TC/SC Secretariats held. Also, eight members are elected from full member NCs, according to the best personal qualifications, balanced geographical distribution, and number of TC/SC Secretariats by their respective NC. The last member is IEC General Secretary, who has a position of ex officio member without vote. The SMB sets up and disbands TCs and SC, as well as approval of their areas of activity (scope). SMB also appoints TC/SC Chairs and allocates secretariats and standards work. When SMB approves it, the TC forms one or more SCs, depending on the needs and extent of the defined goals and work program. Each SC defines its scope under the parent TC. The TC and related SC prepare documents in the defined time and scope. They submit them for voting and approval as International Standards to the full member NCs. Standardization Management Board can establish project committees to prepare specific standards that are outside the scope of an existing TC or SC. The SMB defines timeliness of standards production, reviews, and plans for IEC new work in innovative technological fields. SMB maintains liaisons with other international organizations. It has three subgroups: advisory committees, strategic groups, and ad hoc groups. Advisory committees advise and coordinate IEC work with an ensuring consistency. Strategic groups provide strategic road maps on specific technical areas, especially in new initiatives. Ad hoc groups deal with specific technical topics in a short time (usually less than a year). SMB meets three times in a year.

A *Conformity Assessment* (CA) is any activity that demonstrates correspondence of a product, service, process, body, people, or systems (e.g., management systems) to the specified requirements in formal documents or specifications (typically, but not exclusively, standards). The CAB meets at least once a year. It manages and supervises IEC's CA activities, including IEC's conformity assessment policy, creating, modifying, and disbanding conformity assessment systems, monitoring their operation, and examining continued relevance of the whole IEC conformity assessment activities. Thus, CAB represents the whole IEC conformity assessment community. It has the decision power and reports to IEC CB. The CAB Chair is an IEC Vice-President, elected by the Council. Members are chairs of each conformity assessment system and IEC stand-alone scheme and 15 individual members appointed by the Council. The members without a vote are IEC Treasurer, IEC General Secretary, and secretary of each conformity assessment system and IEC stand-alone scheme.

The Market Strategy Board (MSB) identifies principal technological priorities and market needs for IEC technical and CA work, including response to innovative and fast-moving markets. Sometimes MSB sets up Special Working Group (SWG) for the study of a certain subject and/or to develop a specialized document. Chair, 15 top-level industry technology officers, and IEC officers are members of the MSB, which report to the CB.

The titles and abbreviations of IEC Technical Advisory Committees are given in Table 6.14.

The three IEC SMB Strategic Groups: SG11, SG12, and SG13 are given in Table 6.15.

The list of six IEC Committees is given in Table 6.16.

Table 6.14 IEC Technical Advisory Committees

Committee	Title description
ACEA	Advisory Committee on Environmental Aspects
ACEC	Advisory Committee on Electromagnetic Compatibility
ACEE	Advisory Committee on Energy Efficiency
ACOS	Advisory Committee on Safety
ACSEC	Advisory Committee on Information security and data privacy
ACTAD	Advisory Committee on Electricity Transmission and Distribution

Table 6.15 IEC SMB Strategic Groups

Committee	Description
SG 11	Hot topic radar
SG 12	Digital transformation
SG 13	Working with consortia

Table 6.16 IEC Committees

Committee	Description
EXCO	Executive Committee
FinCom	Finance Committee
IEC-CO	IEC Central Office
ITAG	Information Technology Advisory Group
NRG	New Revenue Generation (NRG)
SAG	Sales Advisory Group

6.2.3 ITU

International Telecommunication Union (ITU) [5] is the United Nations agency for information and communication technologies (ICT) founded in 1865 to facilitate cooperation between international telegraphy networks. Among other things, ITU enabled standardization of the use of Morse code and the first wireless and fixed telecommunications networks.

The most important vision of ITU is to connect the world via ICT that is nowadays a carrier of the modern private and professional life. ITU is responsible for brokering agreement on new technologies and services to be introduced. Since ITU's goal is to create seamless, robust, and reliable global communications systems, its tasks are in the sector of allocation of radio-frequency spectrum and satellite orbital positions.

Topics of ICT include: accessibility, artificial intelligence, climate change and e-waste, consumer information and protection, cybersecurity, digital inclusion, eHealth, emergency communications, gender, human exposure to electromagnetic fields, ICT infrastructure and technologies, innovation, Internet of Things, regulatory and market environments, smart cities and societies, space research, and satellite communications.

ITU's membership consists of diversity of public and private institutions. Members are more than 800 private sector entities and academic institutions from 193 countries. Within the industry, the sector diversity is evident by the presence of small SMEs trying to introduce innovative emerging technologies up to the global largest telecom operators and manufacturers. Therefore, ITU is the important player for decisions on future trends of ICT development.

ITU has in its organization ITU Plenipotentiary Conference, ITU Council, ITU General Secretariat, ITU Telecom, and three main areas of activity organized in Sectors, which work through conferences and meetings (Fig. 6.9).

ITU Plenipotentiary Conference makes ITU top policy-making body with supreme decision power, where its 193 Member States give their final agreement to the ITU's strategic and financial plans, as well as to the leadership for the coming 4-year period. It can also consider and, if required, amend ITU basic texts, like "ITU Constitution and Convention" and "General Rules of Conferences."

ITU Council is a governing body between the Plenipotentiary Conferences. It ensures that ITU's policy is correspondent to the rapidly changing ICT and telecom environment by preparing reports on the policy and strategic planning of the ITU. It facilitates implementation of the provisions of the ITU Constitution, the ITU Convention, and the Administrative Regulations (International Telecommunications Regulations and Radio Regulations [64]), as well as the decisions of Plenipotentiary Conferences. ITU Council also coordinates work programs, approves budgets, and controls finances and expenditure.

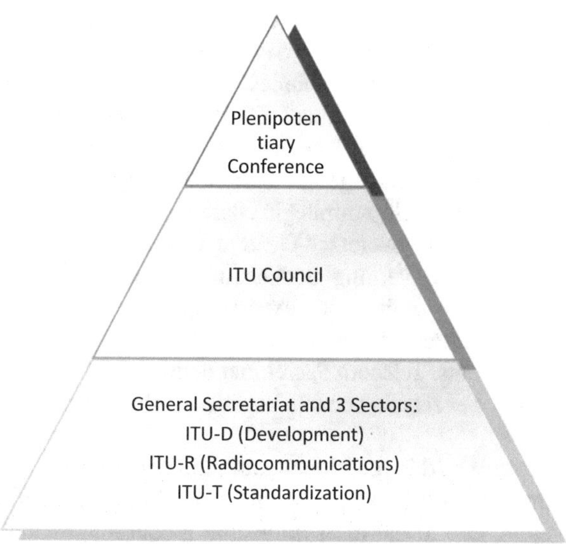

Fig. 6.9 Organization of ITU structure

As of February 2020, for the period 2019–2022, The ITU Council Membership is as follows:

- Region A (the Americas): 9 seats

  Argentina, Bahamas, Brazil, Canada, Cuba, El Salvador, Mexico, United States, Paraguay
- Region B (Western Europe): 8 seats

  France, Italy, Germany, Greece, Hungary, Spain, Switzerland, Turkey
- Region C (Eastern Europe and Northern Asia): 5 seats

  Azerbaijan, Czech Republic, Poland, Romania, Russian Federation
- Region D (Africa): 13 seats

  Algeria, Burkina Faso, Côte d'Ivoire, Egypt, Ghana, Kenya, Morocco, Nigeria, Rwanda, Senegal, South Africa, Tunisia, Uganda
- Region E (Asia and Australasia): 13 seats

  Australia, China, India, Indonesia, Iran (Islamic Republic of), Japan, Korea (Republic of), Kuwait, Pakistan, Philippines, Saudi Arabia, Thailand, United Arab Emirates

ITU Council Troika was established in 2001 pursuant to the recommendation of the Council Working Group on Reform and its Council Resolution 1181 (Annex A) [65]. It consists of the past, current, and future Council Chair. Troika's mandate is "to provide for a smoother transition and continuity in chairing the Council" and to "liaise with the Secretariat to ensure effective preparation for Council sessions."

ITU General Secretariat departments are Office of the Secretary-General (OSG), Strategic Planning and Membership (SPM), Human Resources Management Department (HRMD), Financial Resources Management Department (FRMD), Conferences and Publications Department (C&P), Information Services (IS) department, and ITU Telecom Secretariat. The Secretary-General acts as a legal representative of ITU and holds the OSG together with the Deputy Secretary-General. SPM department analyzes and develops strategic challenges for ITU management team. HRMD gives advice to the Secretary-General in the area of human resources. FRMD gives advices to governing bodies in all areas of financial matters in ITU. C&P provides ITU with linguistic expertise. IS takes care of ITU technology services, including infrastructure, library, safety, and security of staff and of delegates attending ITU events. Telecom Secretariat takes care of planning and delivery of ITU Telecom unique events that enable knowledge sharing, networking, and strategic debates.

The ITU Telecom World is the global gathering place for the ICT sector, open also to ITU nonmembers. It brings economic growth and social good via accelerated ICT innovation, due to the most influential global representation of governments and industries.

Figure 6.10 shows organization of the General Secretariat and ITU Telecom.

The three ITU sectors are:

- Development (ITU-D)
- Radiocommunications (ITU-R)
- Standardization (ITU-T)

They are carried out by corresponding ITU Bureaus. All three ITU sectors organize their world conferences (ITU-D organizes World Telecommunication

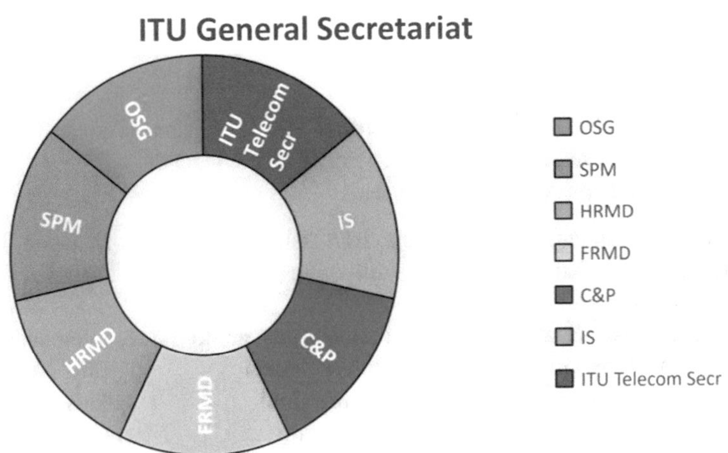

Fig. 6.10 Organization of the General Secretariat and its relation to ITU Telecom

Development Conference (WTDC), ITU-R organizes WRC, and ITU-T organizes World Telecommunication Standardization Assembly (WTSA).

ITU Study Groups (SGs) are focused to define optimum functioning of all services in a specific area with the result of technical standards or guidelines, called Recommendations in ITU.

Except SGs, focus groups work on urgent industry issues outside the mandate of SGs. ITU also organizes seminars on certain topics and workshops that sometimes give outputs to SGs.

ITU Telecommunication Development Sector (ITU-D) has a goal to increase international cooperation and solidarity in technical assistance; to enhance confidence and security in the ICT sector; to enhance environmental protection, climate change adaptation, and disaster management efforts via ICT; and to promote digital inclusion and concentrated assistance to developing countries. ITU-D is the United Nations (UN) specialized agency for implementing projects to enhance ICT development of equipment and networks in developing countries under UN development system or other funding arrangements. As an example, the ITU-D initiative for "bridging the digital divide" is the ITU Connect event.

ITU-D performs its work with the support of the Telecommunication Development Bureau (BDT), the ITU-D Secretariat. BDT has regional offices and four departments: Administration and Operations Coordination Department; Infrastructure, Enabling Environment and E-Applications Department; Innovation and Partnership Department; and Projects and Knowledge Management Department.

ITU-D deals with the following areas: capacity building, climate change and e-Waste, cybersecurity, digital inclusion, emergency telecommunications, ICT applications, innovation, least developed countries, landlocked developing countries and small island developing states, regulatory and market environments, spectrum management and digital broadcasting, statistics and indicators and technology and network development.

ITU with BDT organizes World Telecommunication Development Conference (WTDC) in the period between two Plenipotentiary Conferences to consider related areas relevant to telecommunication development. According to WTDC Resolution 31, one development conference or regional preparatory meeting per region is organized for coordination at the regional level of the preparations for the next World Telecommunication Development Conference (WTDC).

The group that advises the Director of BDT, meets in the interval between WTDCs once a year, and reviews finances, operations, priorities, and strategies of ITU-D is the Telecommunication Development Advisory Group (TDAG).

Structure of ITU-D is shown in Fig. 6.11.

The Global Symposium for Regulators (GSR) is a global annual event of national telecom and ICT regulators all over the world that has the aim of identification of the most urgent regulatory issues in the area. GSR has as a result a set of regulatory best practice guidelines.

ITU also regularly publishes the industry's most comprehensive and reliable ICT statistics, due to ITU BDT's organization of World Telecommunication/ICT Indicators Symposium (WTIS). WTIS is open to whole ITU membership. Usually,

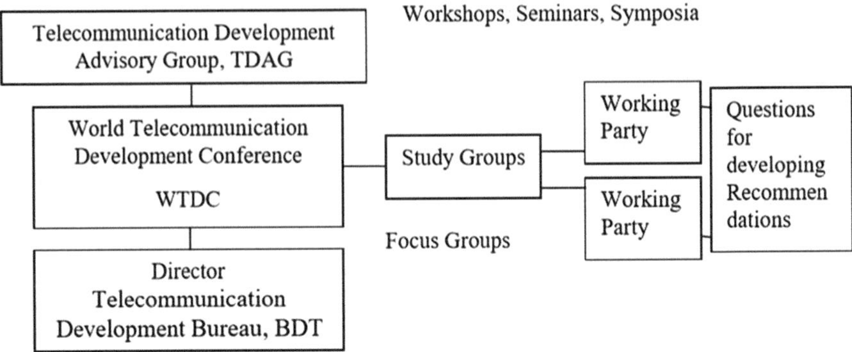

Fig. 6.11 ITU-D structure

a number of 300 delegates participate due to their interest to debate new and emerging issues on ICT data and statistics and their role in policy making and to develop standards and methodologies for producing high-quality data and statistical indicators.

ITU-D SGs develop reports, guidelines, and recommendations from gathered information in surveys, contributions, and case studies. SGs deal with specific task-oriented ICT priorities for developing countries. Initiated policies, project, and strategies strengthen shared knowledge base of Members via online e-Forum, remote participation, or face-to-face meetings.

ITU-D has two SGs:

ITU-D SG1 – Enabling environment for the development of telecommunications/ICTs

ITU-D SG2 – ICT services and applications for the promotion of sustainable development

The mandate of SG1 is related to the national telecom and ICT regulatory, technical, and strategy development to enable benefit from ICT including broadband infrastructure that enables cloud computing, consumer protection, and network functions virtualization. The mandate is also to facilitate implementation of digital economy, as well as to enable ICT access to rural and remote areas and to persons with specific needs and to implement new services.

The mandate of SG2 is devoted to build confidence and security in the use of ICTs and to mitigate impact of climate change. SG2 devotes its efforts to implement conformance and interoperability testing for ICT equipment and to control safe disposal of electronic waste. Finally, the efforts are devoted to control human exposure to electromagnetic fields.

The *ITU Radiocommunication Sector* (ITU-R) deals with the rational, efficient, and economical use of the radio-frequency spectrum by all radiocommunication terrestrial and satellite services. Thus, ITU-R ensures interference-free operations of radiocommunication systems by implementing the Radio Regulations and regional agreements. It allocates bands of radio-frequency spectrum, effects

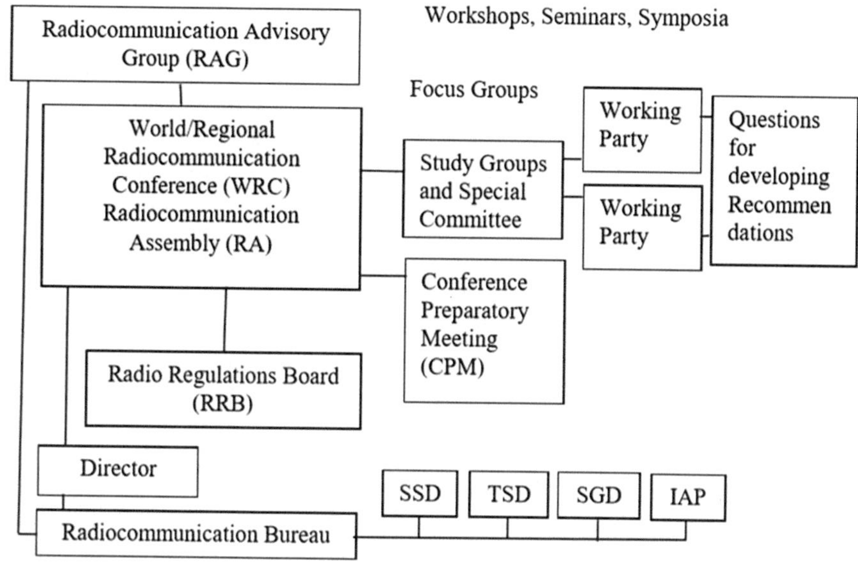

Fig. 6.12 ITU-R structure

allotment of radio frequencies, and registers radio-frequency assignments and any associated orbital position of the geostationary-satellite orbit for avoiding harmful interference between radio stations of various countries.

ITU-R is responsible for studies and recommendations in the area of radiocommunications with the goal of future expansion and new technological development. Structure of ITU-R is shown in Fig. 6.12.

The *World Radiocommunication Conference*, known as WRC, reviews and revises the Radio Regulations, the international treaty governing the use of the radio-frequency spectrum and the geostationary-satellite and non-geostationary-satellite orbits. ITU Council determines whether revisions are required, based on previous recommendations by WRCs 2 years before WRC. WRC can address any radiocommunication matter of worldwide character, instruct the Radio Regulations Board and the Radiocommunication Bureau, and review their activities, revise Radio Regulations and any associated frequency assignment and allotment plans, and determine Questions for study by the Radiocommunication Assembly and its Study Groups in preparation for future Radiocommunication Conferences. All these important functions of the WRC derive from the supreme ITU document, ITU Constitution.

Radio Regulations Board (RRB) has 12 members, elected at the Plenipotentiary Conference. The Executive Secretary of RRB is the Director of ITU-R Bureau. RRB meets up to four times in Geneva per year. RRB advises Radiocommunication Conferences and Radiocommunication Assemblies; approves Rules of Procedure that uses Radiocommunication Bureau when applying provisions of Radio Regulations and registering frequency assignments made by Member States;

formulates Recommendations on the basis of reports of unresolved interference investigations carried out by the Bureau at the request of one or more administrations; considers appeals against decisions made by the Radiocommunication Bureau regarding frequency assignments; takes care of matters unsolvable through application of the Radio Regulations and Rules of Procedure referred by the Bureau; and performs any additional duties of a competent conference or the Council.

Radiocommunication Assembly (RA) is responsible for a program, approval, and structure of radiocommunication studies. RA is usually convened every 3 or 4 years. Sometimes, it may be collocated in time and place with the World Radiocommunication Conferences (WRCs). RA approves and issues ITU-R Recommendations and ITU-R Questions developed by the Study Groups; it sets the program for Study Groups or establishes/disbands certain Study Groups according to the needs; it suggests suitable topics for the agenda of future WRCs; it assigns conference preparatory work for the Study Groups and responds to other requests from ITU conferences.

Conference Preparatory Meeting (CPM) prepares a consolidated report as a support to World and Regional Radiocommunication Conferences in technical, operational, and procedural matters. CPM operates by taking contributions from Radiocommunication Study Groups, administrations, and eventually other sources. CPM usually holds two sessions during the interval between WRCs, mostly for coordinating work programs of the relevant ITU-R Study Groups.

Radiocommunication Bureau deals with the four departments: Space Services Department (SSD), Terrestrial Services Department (TSD), Study Group Department (SGD), and Informatics, Administration and Publications Department (IAP).

In Chap. 4, it is explained that ITU-R publishes ITU-R Recommendations, ITU-R Questions, ITU-R Reports, ITU-R Resolutions, and ITU-R Handbooks. ITU-R publishes also some other documents that are the basis of its work: Radio Regulations (RR) [64], Rules of Procedure [66], and Master International Frequency Register (MIFR) [67].

ITU-R has six Study Groups (SGs), whereby over 5000 specialists deal with the following topics:

ITU-R SG1 – Spectrum Management
ITU-R SG3 – Radio Wave Propagation
ITU-R SG4 – Satellite Services
ITU-R SG5 – Terrestrial Services
ITU-R SG 6 – Broadcasting Services
ITU-R SG 7 – Science Services

ITU Telecommunication Standardization Sector (ITU-T) deals with ITU standards, called ITU-T Recommendations.

Telecommunication Standardization Bureau (TSB) supports the ITU-T SG, as well as ITU-T standards development process in six languages: Arabic, Chinese, English, French, Russian, and Spanish.

The World Telecommunication Standardization Assembly (WTSA) establishes SGs together by appointing chairs and vice-chairs, by meeting every 4 years.

Telecommunication Standardization Advisory Group (TSAG) establishes, restructures, and provides guidelines to SGs. It gives flexibility between WTSAs. TSAG reviews priorities, operations, strategy, and financial matters.

Workshops and seminars are organized for open public by ITU-T to explore new work areas, covering wide ICT areas from various industry sectors.

New ITU-T Recommendations can be a result of a technology watch over emerging technologies and their possible impact on future standardization work in developed and developing countries.

ITU-T drives standards development by membership contributions into a SG, typically by suggesting new work areas, or by suggesting changes to existing Recommendations, or by drafting new Recommendations. WTSA appoints Chair and Vice-chairs of the SGs. Study Questions (SQs) address technical studies of a particular study area, and they are the basic units of the SGs' work. Within SG, Working Party (WP) coordinates several SQs in a certain technical area. A Rapporteur group is a team of experts working on a SQ by studying all relevant Recommendations in order to develop text. Once when the defined work has been completed, SQ is terminated.

As given in Chap. 4, except the ITU-T Recommendations, ITU-T generates documents that serve the membership: TSB Circulars and TSB Collective letters, as well as the meeting documents posted on SGs' web pages (contributions, reports, temporary documents, and liaison statements).

Structure of ITU-T is shown in Fig. 6.13.

ITU-T SGs develop ITU-T Recommendations that define global ICT infrastructure, especially related to ICT interoperability.

The titles of the ITU-T 11 SGs are as follows:

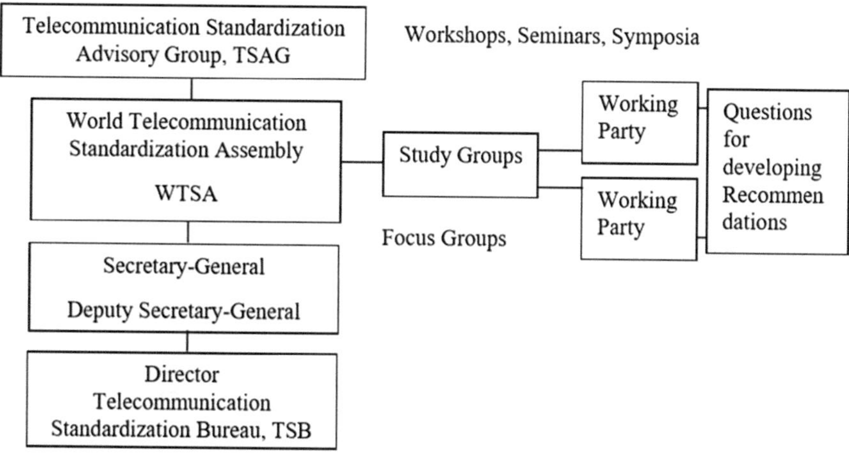

Fig. 6.13 ITU-T structure

ITU-T SG2 – Operational aspects
ITU-T SG3 – Economic and policy issues
ITU-T SG5 – Environment and circular economy
ITU-T SG9 – Broadband cable and TV
ITU-T SG11 – Protocols and test specifications
ITU-T SG12 – Performance, QoS and QoE
ITU-T SG13 – Future networks, with focus on IMT-2020, cloud computing and trusted network infrastructures
ITU-T SG15 – Transport, access and home
ITU-T SG16 – Multimedia
ITU-T SG17 – Security
ITU-T SG20 – IoT, smart cities & communities

Bibliography

1. World Intellectual Property Organisation (WIPO), Internet page: www.wipo.org. Retrieved 17.8.2019
2. C. Binz, B. Truffer, Global innovation systems – a conceptual framework for innovation dynamics in transnational contexts. Research policy. (2017). DOI: https://doi.org/10.1016/j.respol.2017.05.012. Retrieved 1.8.2019
3. International Organization for Standardization, ISO, Internet page: www.iso.org. Retrieved 2.7.2019
4. International Electrotechnical Committee, IEC, Internet page: www.iec.ch. Retrieved 3.8.2019
5. International Telecommunication Union, ITU, Internet page: www.itu.int. Retrieved 1.7.2019
6. World Standards Cooperation, WSC, Internet page: www.worldstandardscooperation.org. Retrieved 2.7.2019
7. United Nations, UN, Internet page: www.un.org/en. Retrieved 2.7.2019
8. Institute of Electrical and Electronics Engineers, IEEE, Internet page: www.ieee.org. Retrieved 3.7.2019
9. Internet Engineering Task Force, IETF, Internet page: www.ietf.org. Retrieved 4.7.2019
10. World Wide Web Consortium, W3C, Internet page: www.w3.org. Retrieved 3.8.2019
11. Universal Postal Union, UPU, Internet page: www.upu.int. Retrieved 8.7.2019
12. Electropedia, Internet page: www.electropedia.org. Retrieved 5.8.2019
13. IEC STATUTES AND RULES OF PROCEDURE (2001 edition, amended on 2004-01-02, 2005-01-07, 2005-09-02, 2006-06-23, 2008-01-18, 2009-02-13, 2011-07-01, 2013-01-11, 2015-01-30, 2017-09-01, 2017-12-15 and 2018-01-12) as endorsed by the National Committees), Internet page: www.iec.ch/members_experts/refdocs/iec/stat_2001-2018e.pdf. Retrieved 4.8.2019
14. IEC GUIDE 103:1980, Internet page: webstore.iec.ch/publication/7515. Retrieved 5.8.2019
15. IEC GUIDE 104:2010, Internet page: webstore.iec.ch/publication/7516. Retrieved 7.8.2019
16. IEC GUIDE 107:2014, Internet page: webstore.iec.ch/publication/7518. Retrieved 3.8.2019
17. IEC GUIDE 108:2006, Internet page: webstore.iec.ch/publication/7519. Retrieved 4.8.2019
18. IEC GUIDE 109:2012, Internet page: webstore.iec.ch/publication/7520. Retrieved 21.7.2019
19. IEC GUIDE 110:2014, Internet page: webstore.iec.ch/publication/7521. Retrieved 22.7.2019
20. IEC GUIDE 111:2004, Internet page: webstore.iec.ch/publication/7522. Retrieved 23.7.2019
21. IEC GUIDE 112:2017, Internet page: webstore.iec.ch/publication/60675. Retrieved 24.7.2019
22. IEC GUIDE 115:2007, Internet page: webstore.iec.ch/publication/7524. Retrieved 25.7.2019
23. IEC GUIDE 116:2018, Internet page: webstore.iec.ch/publication/63780. Retrieved 26.7.2019

24. IEC GUIDE 117:2010, Internet page: webstore.iec.ch/publication/7526. Retrieved 12.8.2019
25. IEC GUIDE 118:2017, Internet page: webstore.iec.ch/publication/33877. Retrieved 14.7.2019
26. IEC GUIDE 119:2017, Internet page: webstore.iec.ch/publication/30779. Retrieved 25.7.2019
27. IEC GUIDE 120:2018, Internet page: webstore.iec.ch/publication/62122. Retrieved 21.8.2019
28. ISO/IEC GUIDE 2:2004 Internet page: webstore.iec.ch/publication/11932. Retrieved 24.7.2019
29. ISO/IEC GUIDE 14:2018, Internet page: webstore.iec.ch/publication/62918. Retrieved 22.8.2019
30. ISO/IEC GUIDE 17:2016, Internet page: webstore.iec.ch/publication/24306. Retrieved 23.7.2019
31. ISO/IEC GUIDE 21-1:2005, Internet page: webstore.iec.ch/publication/11933. Retrieved 24.8.2019
32. ISO/IEC GUIDE 21-2:2005, Internet page: webstore.iec.ch/publication/11934. Retrieved 23.8.2019
33. ISO/IEC GUIDE 23:1982, Internet page: webstore.iec.ch/publication/11935. Retrieved 22.7.2019
34. ISO/IEC GUIDE 37:2012, Internet page: webstore.iec.ch/publication/11937. Retrieved 8.7.2019
35. ISO/IEC GUIDE 41:2018, Internet page: webstore.iec.ch/publication/63937. Retrieved 13.8.2019
36. ISO/IEC GUIDE 46:2017, Internet page: webstore.iec.ch/publication/59996. Retrieved: 11.7.2019
37. ISO/IEC GUIDE 50:2014, Internet page: webstore.iec.ch/publication/11940. Retrieved: 12.8.2019
38. ISO/IEC GUIDE 51:2014, Internet page: webstore.iec.ch/publication/11941. Retrieved: 21.8.2019
39. ISO/IEC GUIDE 59:1994, Internet page: webstore.iec.ch/publication/11943. Retrieved: 24.7.2019
40. ISO/IEC GUIDE 60:2004, Internet page: webstore.iec.ch/publication/11944. Retrieved: 22.6.2019
41. ISO/IEC GUIDE 63:2012, Internet page: webstore.iec.ch/publication/11945. Retrieved: 21.7.2019
42. ISO/IEC GUIDE 68:2002, Internet page: webstore.iec.ch/publication/11947. Retrieved: 25.8.2019
43. ISO/IEC GUIDE 71:2014, Internet page: webstore.iec.ch/publication/11948. Retrieved: 22.6.2019
44. ISO/IEC GUIDE 74:2004, Internet page: webstore.iec.ch/publication/11949. Retrieved: 21.7.2019
45. ISO/IEC GUIDE 75:2006, Internet page: webstore.iec.ch/publication/11950. Retrieved: 14.6.2019
46. ISO/IEC GUIDE 76:2008, Internet page: webstore.iec.ch/publication/11951. Retrieved 15.7.2019
47. ISO/IEC GUIDE 77-1:2008, Internet page: webstore.iec.ch/publication/11953. Retrieved 15.5.2019
48. ISO/IEC GUIDE 77-2:2008, Internet page: webstore.iec.ch/publication/11954. Retrieved 13.6.2019
49. ISO/IEC GUIDE 77-3:2008, Internet page: webstore.iec.ch/publication/11955. Retrieved 14.7.2019
50. ISO/IEC GUIDE 98-1:2009, Internet page: webstore.iec.ch/publication/11956. Retrieved 13.8.2019
51. ISO/IEC GUIDE 98-3:2008, Internet page: webstore.iec.ch/publication/11961. Retrieved 14.8.2019

52. ISO/IEC GUIDE 98-3-SP1:2008, Internet page: webstore.iec.ch/publication/11958. Retrieved 11.6.2019
53. ISO/IEC GUIDE 98-3-SP1:2008/COR1:2009, Internet page: webstore.iec.ch/publication/11959. Retrieved 10.6.2019
54. ISO/IEC GUIDE 98-3-SP2:2011, Internet page: webstore.iec.ch/publication/11960. Retrieved 9.6.2019
55. ISO/IEC GUIDE 98-4:2012, Internet page: webstore.iec.ch/publication/11962. Retrieved 14.6.2019
56. ISO/IEC GUIDE 99:2007, Internet page: webstore.iec.ch/publication/11963. Retrieved 20.7.2019
57. ISO GUIDE 64:2008, Internet page: webstore.iec.ch/publication/9310. Retrieved 20.7.2019
58. ISO GUIDE 72:2001, Internet page: webstore.iec.ch/publication/9311. Retrieved 19.7.2019
59. ISO GUIDE 73:2009, Internet page: webstore.iec.ch/publication/9312. Retrieved 23.7.2019
60. IEC Guide 104 Ed. 6.0 (2010-08): the preparation of safety publications and the use of basic safety publications and group safety publications, Internet page: webstore.iec.ch/publication/7516. Retrieved 6.8.2019
61. IEC Guide 107 Ed. 6.0 (2014-07): electromagnetic compatibility – Guide to the drafting of electromagnetic compatibility publications, Internet page: webstore.iec.ch/publication/7518. Retrieved 7.8.2019
62. IEC Guide 108 Ed. 2.0 (2006-08): guidelines for ensuring the coherency of IEC publications – Application of horizontal standards, Internet page: webstore.iec.ch/publication/7519. Retrieved 5.8.2019
63. World Trade Organization (WTO), www.wto.org. Retrieved 4.8.2019
64. ITU Radio Regulations (RR), Internet page: http://search.itu.int/history/HistoryDigitalCollectionDocLibrary/1.43.48.en.101.pdf. Retrieved 5.8.2019
65. ITU Council Resolution 1181 (Annex A), Internet page: www.itu.int/council/pd/council-resdec.html. Retrieved 6.8.2019
66. ITU Rules of Procedure, Internet page: www.itu.int/en/publications/ITU-R/pages/publications.aspx?parent=R-REG-ROP-2017&media=paper. Retrieved 5.8.2019
67. ITU Master International Frequency Register (MIFR), Internet page: www.itu.int/en/ITU-R/terrestrial/broadcast/Pages/MIFR.aspx. Retrieved 4.8.2019

Chapter 7
Supranational Innovation and Standards Circles

7.1 Supranational Innovation Circle

As in Sects. 3.1 and 6.1, in the supranational innovation circle (SIC), four main stakeholders are involved with an involvement on the supranational level:

- Research and Education Institutions (REI), or Industry (IND) as a nurturing place of inventors
- Governmental and private organizations funding (FUN), forming the entrepreneurial level
- Marketing Organizations (MO), as organized places of marketers
- USers (US) and Users Organizations (UO) of a nongovernmental type

Since the same organization has been already described for the case of global innovation circle and is given in Fig. 6.1, it will not be repeated here. The only change is that instead of the word "global," the word "supranational" has to be applied. However, if the example of the European Union (EU) is taken as a supranational entity, then the role of the EU has to be discussed. The EU is providing funds, which enables REI and IND to grow as places for inventors. The other EU roles include more harmonized protection of Intellectual Property Right (IPR) holders and the lower transaction costs for the content users, as the result of the European regulatory framework for copyright and related rights, discussed in Chap. 4. The framework is the result of inclusive dialogue of all stakeholders on Intellectual Property (IP) issues. The EU is one of the leaders in international negotiations and discussions on patents, copyrights, and related issues.

In SIC, inventors are positioned in the REI or IND. REI is usually organized on the national level, whereas IND is in general a supranational or global organization. Inventors from the both origins (REI and IND) participate in the standardization process. Therefore, SIC contains one more stakeholder, and that is the Supranational Standardization Organization (SSO).

The simplest interplay between five SIC stakeholders can be seen in Fig. 7.1:

D. Šimunić, I. Pavić, *Standards and Innovations in Information Technology and Communications*, https://doi.org/10.1007/978-3-030-44417-4_7

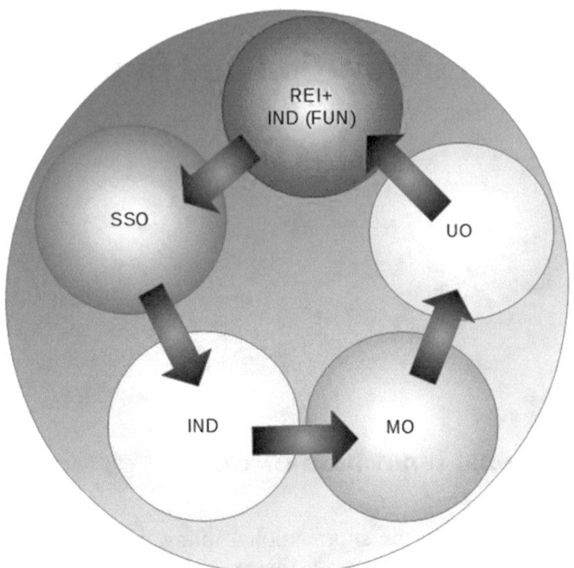

Fig. 7.1 Supranational innovation circle with the five stakeholders

- Research and Education Institutions (REI), or Industry (IND) as a nurturing place of inventors, providing funding (FUN)
- Supranational Standardization Organizations (SSO)
- Industry (IND) producing invention
- Marketing Organizations (MO), as organized places of marketers
- USers (US) and Users Organizations (UO) of a nongovernmental type

If inventors are the IPR holders, SIC stakeholders are:

- Research and Education Institutions (REI) or Industry (IND) (especially small and medium enterprises) as a nurturing place of inventors
- Supranational IP organization (SIPO)
- Supranational Standardization Organizations (SSO)
- Industry (IND) producing invention
- Marketing Organizations (MO), as organized places of marketers
- USers (US) and Users Organizations (UO) of a nongovernmental type

As shown in Fig. 7.2, this case of supranational innovation circle has six stakeholders. As for the global innovation circle, also for the supranational case, a small and medium enterprise (SME) could be a REI with an IPR. In the first phase, REI produces the invention and patents it with the Supranational Intellectual Property Organization, SIPO (an example of a supranational patent organization is EUIPO [1]). As for the global case, after obtaining the patent rights, REI promotes its invention in the SSO and contributes to the new prototype. It is not seldom that the SSO buys the rights itself. Sometimes, another company, which has an interest in

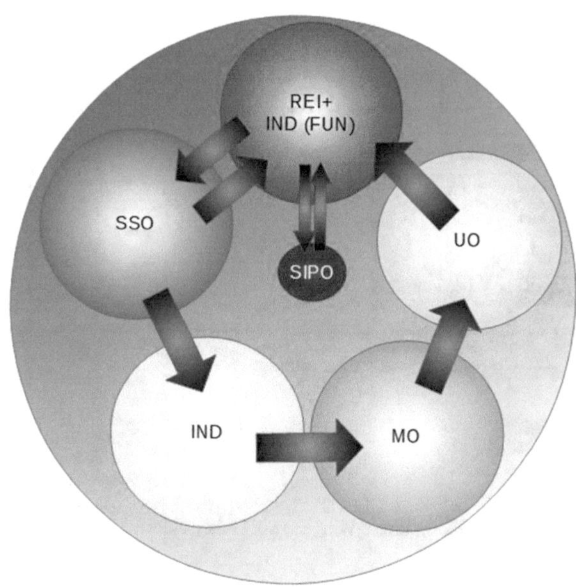

Fig. 7.2 Supranational innovation circle with the six stakeholders

Table 7.1 The four important configuration types of the supranational innovation circle

	SIC type			
Stakeholder	SOSIC	LMSIC	LKSIC	LSIC
REI	SNL	SNL	NL	NL
FUN	SNL	SNL	GL	NL
MO	SNL	NL	SNL	NL
UO	SNL	SNL	SNL	NL
SO	SNL	SNL	SNL	NL, SNL
SIPO	SNL	SNL	SNL	NL, SNL

participating in the valuation, buys the IPR from REI. The rest of the supranational innovation circle is the same: after standardization, industry launches the product, marketers market it, and users buy it and use it. Users in the SIC will give their opinion to REIs and IND how to improve or change it, which enables closing the circle.

There are many possible different configuration types of the supranational innovation circle. However, the most important are four cases, shown in Table 7.1 and in Fig. 7.3. The abbreviations related to the configuration types are the same as in Fig. 7.1, i.e., REI stands for Research and Education Institutions, FUN stands for Governmental and private organizations funding, MO stands for Marketing Organizations, UO stands for Users Organizations of a nongovernmental type, SO stands for Standard Organization, and SIPO stands for Supranational Intellectual Property Organization. Furthermore, "GL" means "Global Level," "SNL" means "SupraNational Level," and "NL" means "National Level."

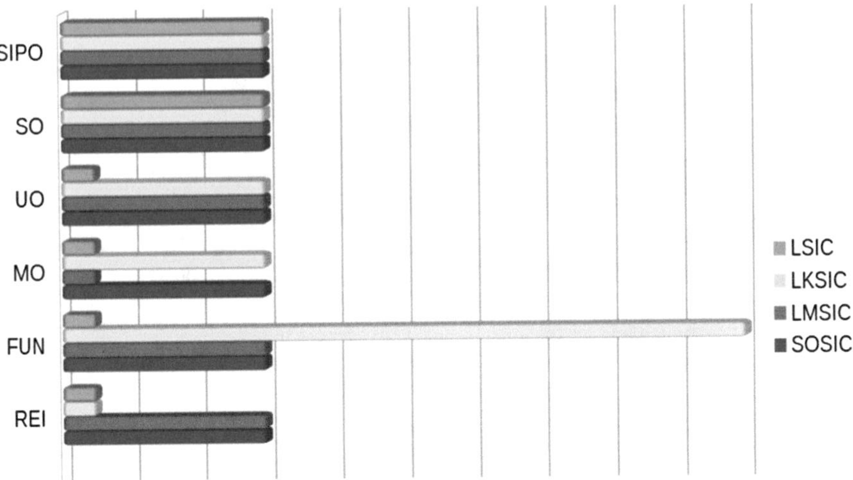

Fig. 7.3 Supranational innovation circle with the six stakeholders

As shown in Table 7.1 and Fig. 7.3, Supranational Oriented SIC (SOSIC) is supranational, meaning that the knowledge is contained in supranational communities (on the SupraNational Level, SNL), funding is of the supranational type, the market is oriented toward the complete price competition, and the nature of users is supranational. Finally, the involved standardization organization and intellectual property organization are both supranational, as well. The second type, Local Market SIC (LMSIC), has the same characteristics of supranational knowledge and funding as the previously discussed SOSIC, but the market has the local nature (i.e., it is on the National Level, NL). The standardization and intellectual property organizations as well as the users are of the supranational nature (i.e., SupraNational Level, SNL). The third type, Local Knowledge SIC (LKSIC), has a market, users, and standardization and intellectual property organization of supranational nature, funding of the global, and knowledge of national nature. Finally, the fourth type, Local SIC (LSIC), is characterized by four characteristics of the local nature (knowledge, funding, market, and users) and by the fact that the standardization and intellectual property organization are or supranational nature.

Figure 7.3 illustrates Table 7.1 more visibly.

7.2 Supranational Standards Circle

The supranational standard circle (SSC) consists of four main stakeholders:

- Supranational Standards Body
- Authority
- Industry
- Societal stakeholders

Supranational Standards Body (SSB) has a central position in the standardization stakeholders circle. If Europe is taken as an example of a supranational entity, then the SSB represents the European Standards Organization (ESO).

An authority is the next important stakeholder in the SSC. Supranational authority is a body with the legal powers and rights, acting on a supranational level. If Europe is taken as an example of a supranational entity, then the supranational authority is the European Union (EU).

Industry is of a supranational or global type, representing all producers, sellers, companies, business and industry associations, foundations, professional bodies, trade associations, etc. Industry has an economic and business interest: to finance the production and to get income out of the production and selling of the product that is a result of the standard.

Societal stakeholders in the SSC are trade unions, supranational consumer organizations, and environmental organizations. They ensure relevance of developed goods and services to market expectations regarding environmental protection. An example of operation of supranational societal stakeholders in the European Union is given in Sect. 3.2.

This chapter allows more introspect into the most important stakeholder, SSB.

The main characteristic of supranational organizations is that they are relevant on the supranational level. In the European Union (EU), according to the "New Approach" system and defined by the Directive 98/34/EC [2], there are three recognized European Standards Organizations (ESO) that can publish the official EU standards: CEN [3], CENELEC [4], and ETSI [5].

7.2.1 CEN

The European Committee for Standardization, CEN [3], is a nonprofit technical organization that was founded according to the Belgian law in 1961. According to the Directive 98/34/EC [2], CEN is the only confirmed European organization for planning, writing, and adopting the European standards in all areas of an economic activity, except for the electrical engineering area. The ESO that is responsible for electrical engineering is CENELEC [4]. The ESO for telecommunications is ETSI [5]. CEN is a part of the European Standardization System, ESS. CEN has an international, multisectoral, and highly decentralized character. The CEN used to have the management center CMC in Brussels, Belgium. Nowadays, CEN and CENELEC joined their management centers in the CEN-CENELEC Management Centre (CCMC) [6] that is in Brussels. CCMC oversees the daily operations, coordination, and promotion of all CEN and CENELEC activities.

CEN is based on the representation of 34 European National Standards Bodies, with 28 EU Members, the Republic of North Macedonia, Serbia, Turkey, and 3 EFTA countries (Iceland, Norway, and Switzerland). European standards are based on a consensus that reflects socioeconomic interests of all 34 CEN countries over their National Standards Bodies (NSBs). It is important to state that most of the standards start with industry initiatives, while a minor part stems from initiatives of consumers, small and big companies, or associations.

CEN network is organized via:

- National Members
- Affiliates
- Associated Bodies
- Partner Organizations
- Liaison Organizations
- European Counsellors
- European Institutional Stakeholders
- Companion Standardization Bodies
- Other Partner Organizations

As of February 2020, 34 National Members are the final decision-makers within CEN (given in Table 7.2).

CEN has three Affiliates, which are NSBs of the countries formally recognized by the European Union as being potential candidates for EU membership. These are the Institute for Standardization of Bosnia and Herzegovina, BAS [7]; General Directorate of Standardization-Albania, DPS [8]; and Institute for Standardization of Montenegro, ISME [9].

CEN has one Associated Body, a European Partner with recognized technical expertise in a specific field that is willing and able to provide preparatory standardization work on a systematic and organized basis to be introduced into the normal CEN standard-making process: The Standardization Association of the European Associations of Aerospace Industries, ASD-STAN [10].

CEN has nine Partner Organizations that have an interest in the cooperation at overall corporate and technical level with CEN. Examples are pan-European professional associations in seven different areas: building industry, building materials, consumers, environment, unions, medical technologies, and small- and medium-sized enterprises: the European Association for the Co-ordination of Consumer Representation in Standardisation, ANEC [11]; European Environmental Citizens Organisation for Standardisation, ECOS; European Trade Union Confederation, ETUC [12]; European Trade Union Institute, ETUI [13]; European Trade Association for the Fire Safety and Security Industry, EURALARM [14]; European Construction Industry Federation, FIEC [15]; Europe's Technology Industries, ORGALIM [16]; Alliance of European medical technology industry association, MedTechEurope [17]; and Small Business Standards, SBS [18].

In February 2020, CEN has 266 Liaison Organizations, i.e., the European organizations that represent interest groups committed to provide input to the work of one or more CEN Technical Bodies. Examples are European Automobile Manufacturers' Association [19], The European Cement Association [20], European Disability Forum [21], European Emergency Number Association [22], European e-Invoicing Service Providers Association [23], European Federation for Elevator Small and Medium-sized Enterprises [24], European Heating Industry [25], European Partnership for Energy and the Environment [26], European Security

Table 7.2 CEN members

National Standards Body	Acronym	Country
Austrian Standards International – Standardization and Innovation	ASI	Austria
Bureau of Normalization/Bureau voor Normalisatie	NBN	Belgium
Bulgarian Institute for Standardization	BDS	Bulgaria
Croatian Standards Institute	HZN	Croatia
Cyprus Organisation for Standardisation	CYS	Cyprus
Czech Office for Standards, Metrology and Testing	UNMZ	Czech Republic
Dansk Standard	DS	Denmark
Estonian Centre for Standardisation	EVS	Estonia
Suomen Standardisoimisliitto r.y.	SFS	Finland
Association Française de Normalisation	AFNOR	France
Deutsches Institut für Normung	DIN	Germany
National Quality Infrastructure System	NQIS/ ELOT	Greece
Hungarian Standards Institution	MSZT	Hungary
Icelandic Standards	IST	Iceland
National Standards Authority of Ireland	NSAI	Ireland
Ente Nazionale Italiano di Unificazione	UNI	Italy
Latvian Standard Ltd.	LVS	Latvia
Lithuanian Standards Board	LST	Lithuania
Organisme Luxembourgeois de Normalisation	ILNAS	Luxembourg
The Malta Competition and Consumer Affairs Authority	MCCAA	Malta
Nederlands Normalisatie-instituut	NEN	Netherlands
Standards Norway	SN	Norway
Polish Committee for Standardization	PKN	Poland
Instituto Português da Qualidade	IPQ	Portugal
Standardization Institute of the Republic of North Macedonia	ISRSM	Republic of North Macedonia
Romanian Standards Association	ASRO	Romania
Institute for Standardization of Serbia	ISS	Serbia
Slovak Office of Standards Metrology and Testing	UNMS SR	Slovakia
Slovenian Institute for Standardization	SIST	Slovenia
Asociación Española de Normalización	UNE	Spain
Swedish Standards Institute	SIS	Sweden
Schweizerische Normen-Vereinigung	SNV	Switzerland
Turkish Standards Institution	TSE	Turkey
British Standards Institution	BSI	United Kingdom

Systems Association [27], European Solar Thermal Industry Federation [28], the European e-Skills Association [29], Near Field Communications Forum [30], Nanotechnology Industries Association [31], the European Security in Health Data Exchange [32], etc. The full list is in Table 7.3.

Table 7.3 with the provided complete list of CEN Liaison Organizations, covering all important areas of the citizens, from European Disability Forum through European Institute for Wood Preservation to Nanotechnology Industries Association, shows the significance of CEN.

CEN has two European Counsellors: one from the European Commission and the other from the EFTA Secretariat.

CEN has five European Institutional Stakeholders. They are relevant European Commission's Agencies, Research Services, and other European intergovernmental organizations. The current list, as of 2020, is as follows: European Commission – Joint Research Centre, EC-JRC; European Defense Agency, EDA; European Union Agency for Network and Information Security, ENISA; European Railway Agency, ERA; and FRONTEX.

CEN has a special arrangement with the NSB that is a member or corresponding member of the International Organization for Standardization (ISO). CEN gives to such an NSB a status of Companion Standardization Body (CSB), so that it can actively participate in CEN's technical harmonization. Table 7.4 shows 17 CEN CSBs, as of February 2020.

CEN has specific partnership agreements with other partner organizations when the organization in question does not fall under any before defined category. There are 17 Other Partner Organizations: International Commission on Illumination, CIE [33]; European co-operation for Accreditation, EA [34]; European cyber security organization ASBL, ECSO; European Cooperation for Space Standardization, ECSS [35]; European Network of Transmission System Operators for Electricity, ENTSO-E [36]; European Network of Transmission System Operators for Gas, ENTSOG [37]; European Patent Organisation, EPO; European Association of National Metrology Institutes, EURAMET [38]; European Organization for Civil Aviation Equipment, EUROCAE [39]; International Federation for Structural Concrete, FIB [40]; International Federation of Standards Users, IFAN [41]; International Telecommunication Union, ITU [42]; NATO Standardization Office, NSO [43]; International Organization of Legal Metrology, OIML [44]; International Union of Railways, UIC; Universal Postal Union, UPU [45]; and ZigBee Alliance [46].

CEN Structure

The CEN structure bases itself on three pillars:

- General Assembly
- Administrative Board
- Presidential Committee

The General Assembly (CEN/AG) is the CEN supreme body. CEN/AG determines the CEN policy and as such consists of the NSB delegates of all CEN members and of selected CEN partners, who attend the AG as observers (such as Affiliates, Companion Standardization Bodies, CENELEC, ETSI, ISO, European

Table 7.3 List of CEN liaison organizations

Organization	Acronym
Alliance for Beverage Cartons and Environment	ACE
Architects' Council of Europe	ACE-CAE
European Automobile Manufacturers' Association	ACEA
Association des Constructeurs Européens de Motocycles	ACEM
Association for Emissions Control by Catalyst AISBL	AECC
European Control Manufacturers Association	AFECOR
AGE Platform Europe	AGE
Association of Issuing Bodies	AIB
Advancing Identification Matters Europe	AIMEU
International Association for Soaps, Detergents and Maintenance Products	AISE
Association of European Producers of Steel for Packaging	APEAL
Home Appliance Europe	APPLiA
Association Européenne des Fabricants de Compteurs d'Eau et de Compteurs d'Energie Thermique	AQUA
AQUA Europa	AQUA Europa
The European Federation of Associations of Locks & Builders Hardware Manufacturers	ARGE
AeroSpace and Defence Industries Association in Europe	ASD
Association of European Refrigeration Compressor Manufacturers	ASERCOM
ASIS International	ASIS International
All Terrain Vehicle Industry European Association	ATVEA
Biobanking and Biomolecular Resources Research Infrastructure – European Research Infrastructure Consortium	BBMRI-ERIC
Baby Carrier Industry Alliance	BCIA
International Bureau for Precast Concrete	BIBM
European Federation of Insurance Intermediaries	BIPAR
CANCER-ID	CANCER-ID
Association of Chocolate, Biscuit and Confectionery Industries of Europe	CAOBISCO
Committee for European Construction Equipment	CECE
Committee of European Manufacturers of Petroleum Measuring and Distributing Equipment	CECOD
Council of European Dentists	CED
European Chemical Industry Council	CEFIC
European Committee of Sugar Manufacturers	CEFS
European Confederation of Woodworking Industries	CEI-Bois
Comité Européen de l'Industrie de la Robinetterie	CEIR
European Committee of Associations of Manufacturers of Agricultural Machinery	CEMA
The European Cement Association	CEMBUREAU
Confederation of Inspection and Certification Organisations	CEOC International

(continued)

Table 7.3 (continued)

Organization	Acronym
European Confederation of Paint, Printing Ink and Artists' Colours Industry	CEPE
Confederation of European Paper Industries	CEPI
Council of European Professional Informatics Societies	CEPIS
The European Ceramic Industry Association	CERAME-UNIE
International Committee on Industrial Chimneys	CICIND
European Man-made Fibres Association	CIRFS
European Association for Forwarding, Transport, Logistic and Customs Services	CLECAT
European Association of Automotive Suppliers	CLEPA
Environmental Science for the European Refining Industry	CONCAWE
Pan-European Confederation of Customs Brokers and Custom Representatives	CONFIAD
Centre de Coopération pour les Recherches Scientifiques Relatives au Tabac	CORESTA
Construction Products Europe	CPE
Confederation of European Security Services	CoESS
ComMUnion Project	ComMUnion Project
The Personal Care Association	Cosmetics Europe
DLMS User Association	DLMS UA
Driving Innovation in Crisis Management for European Resilience	DRIVER+ Project
Digital Trust and Compliance Europe	DTCE
DIGITALEUROPE	DIGITALEUROPE
European Autoclaved Aerated Concrete Association	EAACA
European Association for External Thermal Insulation Composite Systems	EAE
European Asphalt Pavement Association	EAPA
European Association for Passive Fire Protection	EAPFP
European Association for the Streamlining of Energy Exchange-gas	EASEE-gas
European Biogas Association	EBA
European Biodiesel Board	EBB
European Biostimulants Industry Council	EBIC
European Balloon & Party Council	EBPC
European Chimneys Association	ECA
European Cocoa Association aisbl	ECA
European Cockpit Association	ECA
European Casino Association	ECA
European Consortium of Anchors Producers	ECAP
European Coil Coating Association	ECCA
European Convention of Constructional Steelwork Associations	ECCS
European Cyclists' Federation	ECF
European Copper Institute	ECI
European Cellulose Insulation Association	ECIA

(continued)

Table 7.3 (continued)

Organization	Acronym
European Cylinder Makers Association	ECMA
European Compost Network ECN e.V.	ECN e.V.
ECO -Platform AISBL	ECO -Platform AISBL
European Coal Combustion Products Association e.V.	ECOBA
European Concrete Platform	ECP
European Cool Roofs Council	ECRC
European Calcium Silicate Producers Association	ECSPA
International Association Serving the Nonwovens and Related Industries	EDANA
European Association for Directors and Providers of Long-Term Care Services for the Elderly	EDE
European Disability Forum	EDF
The Global Network for B2B Integration in High Tech Industries	EDIFICE
European Door and Shutter Federation e.V.	EDSF
European Emergency Number Association	EENA
European e-Invoicing Service Providers Association	EESPA
European Fertiliser Blenders Association	EFBA
European Federation of Engineering Consultancy Associations	EFCA
European Federation of Concrete Admixtures Associations Limited	EFCA
European Federation of Campingsite Organisations & Holiday Park Associations	EFCO&HPA
European Federation for Elevator Small and Medium-sized Enterprises	EFESME
European Federation of Food, Agriculture and Tourism Trade Unions	EFFAT
European Federation of Funeral Services	EFFS
European Federation of Clinical Chemistry and Laboratory Medicine	EFLM
European Federation for Non-Destructive Testing	EFNDT
European Federation for Services to Individuals	EFSI
European Fire Sprinklers Network	EFSN
European Federation for Cosmetic Ingredients	EFfCI
European Gaming and Betting Association	EGBA
European Garage Equipment Association	EGEA
European General Galvanizers Association	EGGA
European Garden Machinery Industry Federation	EGMF
European Group of Organisations for Fire Testing, Inspection and Certification	EGOLF
European Heating Industry	EHI
Euroheat & Power	EHP
European Industrial Gases Association	EIGA
European Lotteries Association	EL
European Lift Association	ELA
European Lift & Lift Components Association	ELCA
European Manufacturers of Feed Minerals Association	EMFEMA

(continued)

Table 7.3 (continued)

Organization	Acronym
European Mortar Industry Organisation	EMO
Vereniging EN 13606 Consortium	EN13606
European Network for Accessible Tourism	ENAT
European Network of Forensic Science Institutes	ENFSI
European Nursery Products Confederation	ENPC
European Organization for Quality	EOQ
European Association for Professions in Biomedical Science	EPBS
European Partnership for Energy and the Environment	EPEE
European Phenolic Foam Association	EPFA
European Perimeter Protection Association	EPPA
European Rental Association	ERA
European Union Road Federation	ERF
European Racking Federation	ERF
European Ready-Mixed Concrete Organisation	ERMCO
European Recovered Paper Association	ERPA
European Sealing Association	ESA
European Society of Aerospace Medicine	ESAM
European Society of Pathology	ESP
European Security Systems Association	ESSA
European Specialist Sports Nutrition Alliance	ESSNA
European Solar Thermal Industry Federation	ESTIF
European Synthetic Turf Organisation	ESTO
European Society of Tattoo and Pigment Research	ESTP
European Single ply Waterproof Association	ESWA
European Transport Workers' Federation	ETF
European Tyre & Rubber Manufacturers' Association	ETRMA
European Textile Services Association	ETSA
European Bentonite Association	EUBA
European Committee of Woodworking Machine Manufacturers	EUMABOIS
European Manufacturers of Expanded Polystyrene	EUMEPS
European Apparel and Textile Organization	EURATEX
European Insulation Manufacturers Association	EURIMA
European Association of Air Heater Manufacturers	EURO-AIR
European Bitumen Association	EUROBITUME
European Steel Association AISBL	EUROFER AISBL
European Committee of the Manufacturers of Fire Engines and Apparatus	EUROFEU
Association of European Gypsum Industries	EUROGYPSUM
European Group for Rooflights and Smoke Ventilation	EUROLUX
European Association of Mining Industries, Metal Ores & Industrial Minerals	EUROMINES
European Association of Internal Combustion Engine Manufacturers	EUROMOT

(continued)

Table 7.3 (continued)

Organization	Acronym
European Association for Bioindustries	EUROPABio
European Organization for Packaging and the Environment aisbl	EUROPEN
European Committee of Pump Manufacturers	EUROPUMP
European Slag Association	EUROSLAG
Europe's Industry Association for Indoor Climate, Process Cooling, and Food Cold Chain Technologies	EUROVENT
European Union for Swimming Pool and Spa Associations	EUSA
European Vending Association	EVA
European Ventilation Industry Association	EVIA
European Water Association	EWA
European Waterproofing Association AISBL	EWA Europe
European Federation for Welding, Joining and Cutting	EWF
European Writing Instruments Manufacturer's Association	EWIMA
European Wood Preservative Manufacturers Group	EWPM
European Water Treatment Association	EWTA
European Expanded Clay Association	EXCA
European e-Skills Association	EeSA
European Composites Industry Association	EuCIA
European Lime Association AISBL	EuLA
European Plastics Converters	EuPC
European Salt Producers' Association	EuSalt
European Cabin Crew Association	EurECCA
European window, door and curtain wall manufacturers	EuroWindoor AISBL
Eurogroup for animals	Eurogroup for animals
EuropeActive	EuropeActive
European Aluminium	European Aluminium (former EAA)
European Bioplastics	European Bioplastics
Federation of European Window and Curtain Wall Manufacturers' Association	FAECF
Association of European manufacturers of Gas Meters, Gas Pressure Regulators and associated Safety Devices and Stations	FARECOGAZ
European Federation of Aerosol	FEA
Fédération Européenne de l'Industrie des Aliments pour Animaux Familiers	FEDIAF
EU Association of Specialty Feed Ingredients and their Mixtures	FEFANA
European Federation of Tourist Guide Associations	FEG
Association of the European Adhesive and Sealant Industry	FEICA
European Federation of Materials Handling and Storage Equipment	FEM
European Envelope Manufacturers' Association	FEPE
European Container Glass Federation	FEVE
Fédération Internationale des Cadres des Transports	FICT
European Dental Industry	FIDE

(continued)

Table 7.3 (continued)

Organization	Acronym
Fédération Internationale de Football Association	FIFA
Fédération Internationale de Motocyclisme	FIM
FluoroCouncil Europe	FluoroCouncil Europe
Confederation of the Food and Drink Industries in the EU	FoodDrinkEurope
Global Cabin Air Quality Executive	GCAQE
European Gas Research Group	GERG
Gas Infrastructure Europe	GIE
GS1	GS1
GS1 in Europe	GS1 in Europe
–The Gaming Standards Association Europe	GSA Europe
Glass for Europe	Glass for Europe
GlobalPlatform	GlobalPlatform
Health Level Seven International Foundation	HL7 International Foundation
Hotels, Restaurants & Cafés in Europe	HOTREC
International Council on Monuments and Sites	ICOMOS
International Dairy Federation	IDF
International Federation of Clinical Chemistry and Laboratory Medicine	IFCC
International Fragrance Association	IFRA
International Guide Dog Federation	IGDF
Global Wallcoverings Association	IGI
International Life Saving Federation of Europe	ILSE
Industrial Minerals Association–Europe	IMA-Europe
International Association of Oil & Gas Producers	IOGP
International Sustainability and Carbon Certification	ISCC
International Society of Hair Restoration Surgery	ISHRS
International Tennis Federation	ITF
International Water Mist Association	IWMA
International Zinc Association – Europe	IZA-Europe
	KRAKEN EU Project
LEADing Practice	LEADing Practice
LightingEurope AISBL	LightingEurope
European LPG Association	Liquid Gas Europe
Technical Association of the European Natural Gas Industry	MARCOGAZ
Methanol Institute	MI
Metals for Buildings asbl	Metals for Buildings
International Natural and Organic Cosmetics Association	NATRUE
Near Field Communication Forum	NFC Forum
Natural & bio Gas Vehicle Association	NGVA Europe
Nickel Institute	NI
Nanotechnology Industries Association	NIA

(continued)

Table 7.3 (continued)

Organization	Acronym
–"Development and implementation of Grouping and Safe-by-Design approaches within regulatory frameworks"	NanoReg2 Project
OpenPEPPOL AISBL	OpenPEPPOL
Paraglider Manufacturers Association	PMA
European Association for Panels and Profiles	PPA Europe
PRE Plastics Recyclers Europe	PRE
Federation of European Polyurethane Rigid Foam Associations	PU Europe
A productive, Affordable and Reliable solution for large scale manufacturing of metallic components by combining laser-based ADDitive and subtractive	Paraddise Project
PlasticsEurope AISBL	PlasticsEurope
Sterile Barrier Association Limited	SBA
SME Safety a.i.s.b.l.	SME Safety
European Security in Health Data Exchange	ShiELD Project
Sustainability Transition Assessment and Research of Bio-based Products	Star-Probio Project
European Plastic Pipes and Fittings Association	TEPPFA
Toy Industries of Europe	TIE
European Livestock and Meat Trading Union	UECBV
European Aggregates Association	UEPG
International Union of Wagon Keepers a.i.s.b.l.	UIP
International Association of Public Transport	UITP
European Rail Industry	UNIFE
Union of European Petroleum Independents	UPEI
VGB PowerTech	VGB
Vacuum Insulation Panel Association	VIPA International
Visa Europe Services INC	VISA EUROPE
Vocational Training Charitable Trust	VTCT
World Association of Manufacturers of Bottles and Teats	WBT
European Institute for Wood Preservation	WEI-IEO
World Federation of the Sporting Goods Industry	WFSGI
Wize Alliance	WIZE
World Rugby Limited	World Rugby
buildingSMART International Ltd	bSI
European Producers Union of Renewable Ethanol	ePURE
European Spirits Organisation	spiritsEUROPE

Partner Organizations, the European Commission, and the EFTA Secretariat). However, voting rights are restricted to the CEN NSBs. The CEN President chairs CEN/AG meetings twice a year. The CEN/AG appoints members of the Administrative Board.

The Administrative Board (CA) is responsible for management and administration of CEN's business, as a support of CEN/AG decisions.

Table 7.4 CEN companion standardization body

National Standards Body	Acronym	Country
National Institute of Standards CJSC	SARM	Armenia
Standards Australia Limited	SA	Australia
Azerbaijan Standardization Institute	AZSTAND	Azerbaijan
State Committee for Standardization of the Republic of Belarus	BELST	Belarus
Agence des Normes et de la Qualité	ANOR	Cameroon
Standards Council of Canada	SCC	Canada
Egyptian Organization for Standardization and Quality	EOS	Egypt
Georgian National Agency for Standards and Metrology	GEOSTM	Georgia
Standards Institution of Israel	SII	Israel
Jordan Standards and Metrology Organization	JSMO	Jordan
Committee for Standardization, Metrology and Certification	KAZMEMST	Kazakhstan
Institute for Standardization of Moldova	ISM	Moldova, Republic of
Mongolian Agency for Standardization and Metrology	MASM	Mongolia
Institut Marocain de Normalisation	IMANOR	Morocco
Standards New Zealand, Ministry of Business, Innovation & Employment	SNZ	New Zealand
National Institute for Standardization and Industrial Property	INNORPI	Tunisia
Ukrainian scientific research and training center of issues of standardization, certification and quality	DSTU	Ukraine

The Presidential Committee (PC) of CEN is a part of the joint structure with the CENELEC Presidential Committee, in order to enable easier cooperation on strategic matters between the two bodies, related to roles of NSBs, the National Electrotechnical Committees (NECs) and the CEN-CENELEC Management Centre (CCMC). The name of the joint structure is the CEN-CENELEC Presidential Committee. Currently, it has 10 members: the two Presidents of CEN and CENELEC, the Presidents-Elect, the six Vice-Presidents, and the Director General of CEN and CENELEC.

Figure 7.4 shows the CEN structure.

The technical bodies that perform the work are:

- Technical Board (BT)
- Technical Bodies (Technical Committees (TCs) and their Sub-Committees (SCs), Working Groups (WGs), BT Task Force (BTTF), and BT Working Groups (BTWGs).

The Technical Board is a body which fully controls and is thus fully responsible for the complete standards program. Technical Committees, the CEN-CENELEC Management Centre (CCMC), and other bodies execute the standards program. BT reports to the General Assembly (CEN/AG). BT responsibilities include decisions on all matters concerning the organizations, coordination and planning of standards work between TC, approving CEN technical policies and strategies, approving TC business plans, taking decisions on standardization issues if required from the TC, and

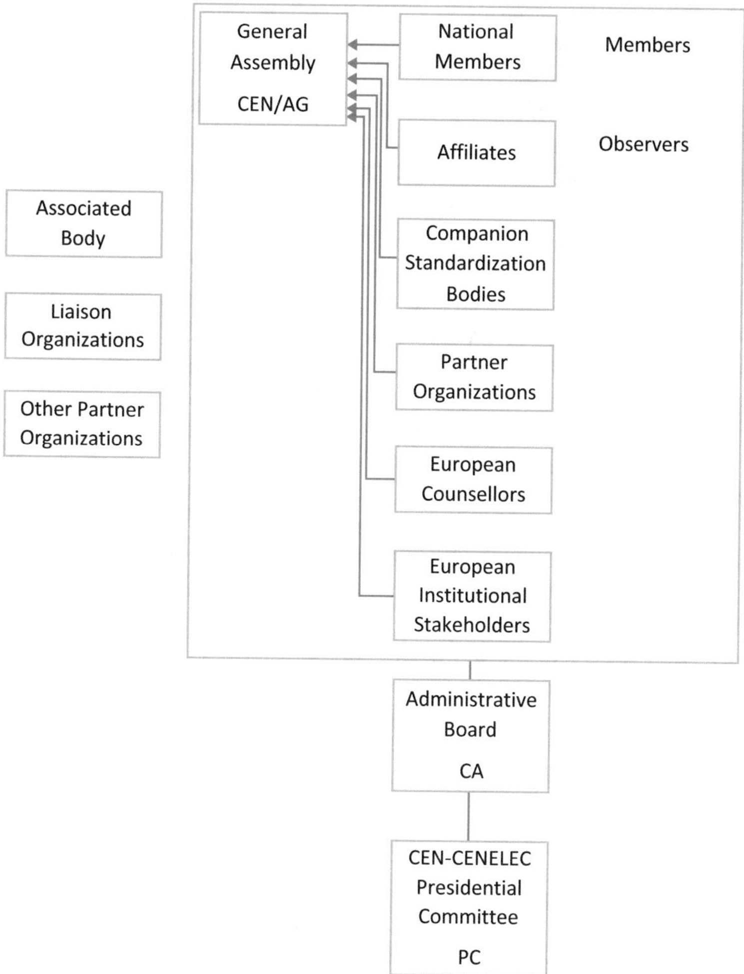

Fig. 7.4 CEN structure

organizing technical liaisons with international organizations, intergovernmental organizations, and European technical, scientific, trade, and professional organizations.

As of February 2020, 399 TCs with 58 SCs and 1621 WGs work with their own scope for development and execution of certain standards. The basis of TC's work is the CEN national participation, which enables a possibility to reach consensus within the whole CEN society. CEN National Members appoint experts, who work in Working Groups and develop the true standards work, which results in the draft and later in the future standard.

In the context of the Regulation 1025/2012 [47], Article 4.4, the CEN National Committees (NCs) can notify their intentions for a work on a new national standard or to revise one of their national standards not already covered by standstill (so-called Vilamoura procedure [48]). New work item proposal can be initiated either via CEN BT or via a corresponding TC, depending on the number of interested NCs. As soon as one other NC indicates interest, the standstill obligation applies, and related TCs and SCs can provide comments. There are several ways of proceeding. The BT decides whether the proposal is relevant for ISO or the work should be started within CEN. If four or more NCs have interest in participation, BT will form a Task Force of the Technical Board (BTTF). BTTF is a technical body that undertakes short-term standardization task of a specific nature in a given time. BTTF consists of a convenor and national delegations.

BT can also ask TC or SC to set up a relevant WG. If less than four NCs are participating, the work will be carried out by the proposing NC or the relevant TC or the SC that will establish a WG. If the only interested NC is the one who proposed the work, it will be carried out just on the level of that country and the standstill will be withdrawn.

CEN Technical Board (CEN BT) sets up a BT Working Group (BTWG) for establishing technical need for information, advice, a study, or rules. CEN BT defines its composition and disbands it after the finished task.

CEN technical structure is shown in Fig. 7.5.

Therefore, the structure of CEN reflects the way of standards development, which is driven as a "bottom-up" approach, i.e., to its relevance of standards, which meets the needs of the market.

As of February 2020, more than 60,000 experts participate in 2134 Technical Bodies: 330 CEN Technical Committees (TCs), 45 Subcommittees of CEN TCs, 1576 CEN Working Groups, and 35 CEN Workshops. Experts cooperate in the 16

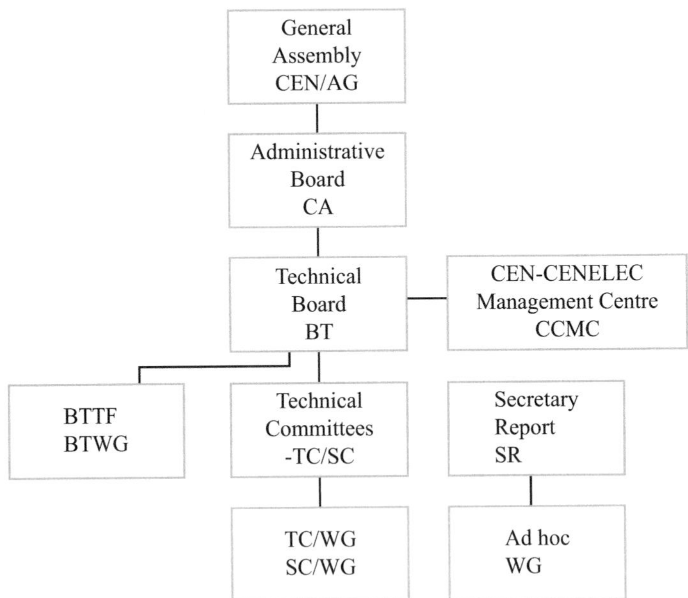

Fig. 7.5 CEN technical structure

CEN-CENELEC TCs, 27 CEN-CENELEC WGs, 6 CEN-CENELEC Workshops, 4 CEN-CENELEC-ETSI TCs, and 11 CEN-CENELEC-ETSI WGs. All the CEN Technical Bodies are working in 22 different activity sectors: Accessibility, Air and space, Bio-based products, Chemistry, Construction, Consumer products, Energy and utilities, Environment, Food, Health and safety, Healthcare, Heating, ventilation and air conditioning (HVAC), ICTs, Innovation, Machinery safety, Materials, Measurement, Nanotechnologies, Pressure equipment, Security and defense, Services and Transport, and packaging.

CEN Deliverables

CEN delivers European Standards (EN), Technical Specifications (TS), Technical Reports (TR), Guides (CG), CEN and/or CENELEC Workshop Agreements (CWA), Pre-standards (ENV), Reports (CR), and European Standards identical to International Standards (ISO). Different market needs drive CEN to produce different kinds of deliverable, related to time required for a development, an approval process, and an implementation. More about CEN deliverables may be found in Chap. 4. The most important CEN deliverable is the European Standard (EN), prepared by a TC. Its operational basis is a national participation of all the CEN members. TC can also create a subcommittee (SG), if needed. Technical standards are mostly developed by a Working Group (WG). EN can be delivered only by achieved full consensus, and it has automatically status of national standard in all CEN Member Countries. CEN TC can use CEN TS as a European Pre-standard (ENV) for some special case of testing new technology before developing the full EN. CEN Workshop Agreement (CWA) is a result of an organized workshop by CEN, when a quick reaction is needed for rapidly changing technology. CWA is open to the direct participation of all interested parties. The publishing time is between 10 and 12 months.

In 2019, there were 1071 European Standards, 50 Technical Specifications, 26 Technical Reports, 15 CEN Workshop Agreements, and 2 CEN Guides (CGs), which makes a total of 1164 produced documents (Fig. 7.6).

The total number of published ENs and other CEN deliverables is 17,309 at the end of 2019. Published are 15,605 European Standards, 26 European Pre-standards, 534 Technical Specifications, 82 CEN Reports (CRs), 546 Technical Reports (TRs), 476 CEN Workshop Agreements, and 40 CEN Guides (CGs) (Fig. 7.7).

European Standards in CEN require ca 800 million EUR, whereas 80% are covered by industry. The DIN study shows that around 1% of Gross National Product is saved due to the standardization in Germany.

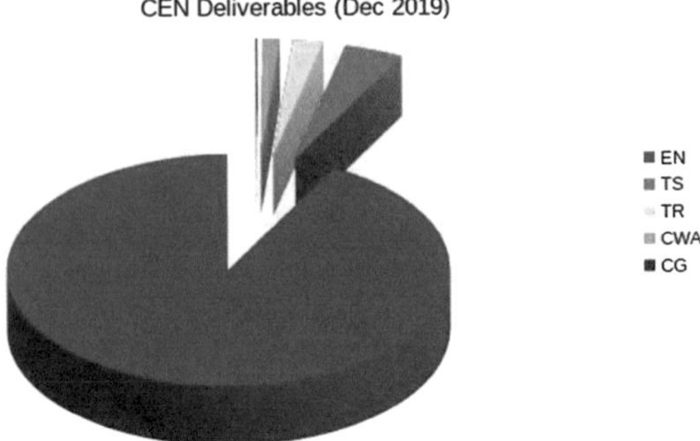

Fig. 7.6 CEN deliverables in December 2019

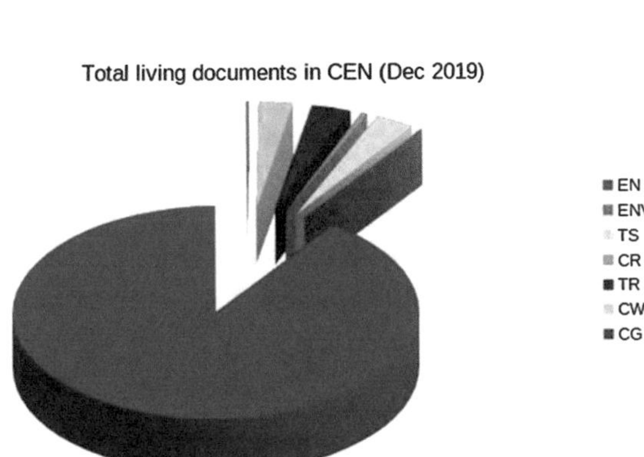

Fig. 7.7 Total CEN deliverables in December 2019

7.2.2 CENELEC

European Committee for Electrotechnical Standardization, CENELEC [4], is a non-profit organization. CENELEC was founded in 1973, by joining two organizations: European Committee for the Coordination of Electrotechnical Standards in the European Economic Community, CENELCOM, and European Committee for the Coordination of Electrical Standards, CENEL. CENELEC is a European supranational Standards Body that is responsible for standardization in the field of electrotechnical engineering. It consists of 34 national electrotechnical committees and 13 third-country "National Electrotechnical Committees" that participate in the work of CENELEC as "Affiliates" and/or "Companion Standardization Bodies" (CSBs) or via Cooperation Agreement. Only the "National Electrotechnical Committees" that are members or associate members of the International Electrotechnical Commission (IEC) can get the status of CSBs. CENELEC cooperates with other supranational standardization bodies via Memorandum of Understanding (MoU). The mission of CENELEC is a preparation of voluntary standards in the field of electrical engineering that are a support to development of united European market for electrical and electronic products and services. In this way, the new markets appear, as well as cut, conformity assessment expenses.

CENELEC technical structure is shown in Fig. 7.8.

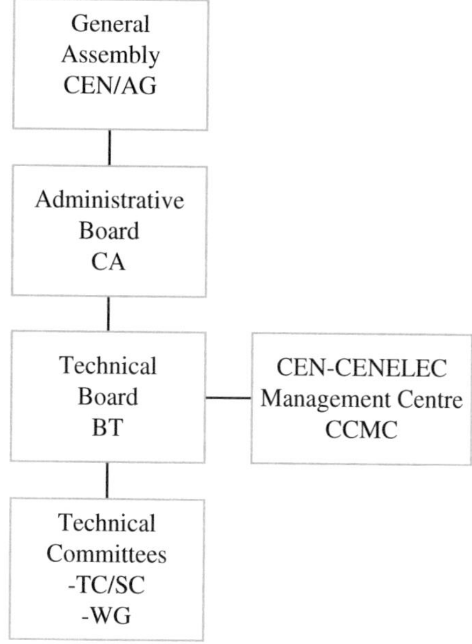

Fig. 7.8 CENELEC technical structure

CENELEC works on improvement of product's and service's quality and security, including environment protection, accessibility and innovation, and gives the support to IEC. The primary focus is oriented toward a security and a free movement of products and services, ensured by European Directives and reached by complete compatibility with IEC International Standards and with suggestions and drafts by CENELEC partners (examples are consumer associations).

One of the important CENELEC's working areas is to foster trade and welfare inside and outside the European Economic Area (performed by publishing European Standards (EN)), and this is exactly the main reason for CENELEC to be open to a collaboration with other standardization bodies worldwide. In 2016 the "Frankfurt agreement" formalized cooperation with IEC that ensures that the majority of the developed CENELEC standards are identical to the IEC standards. The cooperation is enhanced by presence of CENELEC members directly in IEC work and vice versa. Thus, by applying CENELEC standards, the manufacturers have a granted access not only to all 34 CENELEC countries but also worldwide, due to the IEC.

CENELEC standardization work is carried out in different technical bodies in which a wide range of interests is represented. CENELEC Technical Bodies are CENELEC Technical Committee, CENELEC Subcommittee, and Task Force of the Technical Board. Technical Board (BT) establishes a Technical Committee (TC) with exact title and scope to prepare a specific CENELEC publication. CENELEC National Committees' Representatives are members of a TC, and they appoint delegates for their participation at TC meetings. BT appoints TC Chair and allocates TC secretariat to one of the CENELEC members, whose National Committee appoints the TC Secretary. TCs take any IEC work or any data supplied by Member Countries and other relevant international organizations coming within their field of work and put it to the work in the related TCs.

The CENELEC technical bodies consist of national delegates and work on the normative documents: European Standards (ENs), Technical Specifications, etc. They report directly to the CENELEC Technical Board. As of February 2020 (13.2.2019), there are 79 Technical Committees, 15 Subcommittees, 12 BTTF, 4 BTWG, 1 JWG, 78 SR, 10 WS, and 288 TCs and SCs WGs, all active in the following CENELEC sectors: electric vehicles, smart grids, smart metering, household appliances, information and communication technologies (ICTs), electromagnetic compatibility (EMC), electrical engineering, fiber-optic communications, fuel cells, medical equipment, railways, and solar (photovoltaic) electricity systems.

The list of CENELEC Technical Committees, as of April 7, 2019, is given in Table 7.5.

The list of CENELEC SCs is given in Table 7.6.

In the context of the Regulation 1025/2012 [47], Article 4.4, the CENELEC National Committees (NC) can notify their intentions for a work on a new national standard or to revise one of their national standards that is not already covered by the standstill (cf. the already mentioned Vilamoura procedure [48]). A new work item proposal can be initiated either via the CENELEC BT or via a corresponding TC, depending on a number of interested NCs. As soon as one other NC indicates interest, the standstill obligation applies and the related TCs and SCs can provide comments. There are several ways of proceeding. The BT decides whether the proposal is relevant for IEC or the work should be started within CENELEC. If four or more NCs have interest in participation, BT will form Task Force of the Technical Board (BTTF). BTTF is a technical body that undertakes short-term standardization task of a specific nature in the required time period. BTTF consists of a convener and of national delegations.

BT can also ask TC or SC to set up a relevant WG. If less than four NCs are participating, the proposing NC will carry the work or relevant TC or SC will establish a WG. If the only interested NC is the one who proposed the work, it will be carried out just on the level of that country and the standstill will be withdrawn. Table 7.7 gives list of CENELEC BTTF.

CENELEC BT (CENELEC Technical Board) sets up a BT Working Group (BTWG) for establishing technical need for information, advice, a study, or rules. CENELEC BT defines its composition and disbands it after the finished task. As of May 7 2019, the list of CENELEC BTWG is given in Table 7.8.

In addition to CEN and ETSI, CENELEC is the official standardization body of EC. It prepares standards in support to the "New Approach," according to the given mandate of the European Commission. CENELEC is responsible for preparation of standards in support of the Ecodesign Directive (2009/125/EC) [49], Waste from electrical and electronic equipment directive (WEEE) [50]. All the ENs are published in the *Official Journal of the European Communities* [51].

The two most important CENELEC publications are European Standard, EN, and Harmonization Document, HD. Both documents are called "standards," but the difference is that EN has to be taken with the exact wording, and only the technical content is what is important with HD. EN is the most important three-lingual (English, French, and German) CENELEC product, based on consensus, openness, and transparency.

Table 7.5 List of CENELEC Technical Committees

Committee	Title
CLC/TC 100X	Audio, video and multimedia systems and equipment and related sub-systems
CLC/TC 106X	Electromagnetic fields in the human environment
CLC/TC 107X	Process management for avionics
CLC/TC 108X	Safety of electronic equipment within the fields of Audio/Video, Information Technology and Communication Technology
CLC/TC 11	Overhead electrical lines exceeding 1 kV a.c. (1.5 kV d.c.)
CLC/TC 111X	Environment
CLC/TC 116	Safety of motor-operated electric tools
CLC/TC 121A	Low-voltage switchgear and controlgear
CLC/TC 13	Electrical energy measurement and control
CLC/TC 14	Power transformers
CLC/TC 17AC	High-voltage switchgear and controlgear
CLC/TC 18X	Electrical installations of ships and of mobile and fixed offshore units
CLC/TC 2	Rotating machinery
CLC/TC 20	Electric cables
CLC/TC 204	Safety of electrostatic painting and finishing equipment
CLC/TC 205	Home and Building Electronic Systems (HBES)
CLC/TC 209	Cable networks for television signals, sound signals and interactive services
CLC/TC 210	Electromagnetic compatibility (EMC)
CLC/TC 213	Cable management systems
CLC/TC 215	Electrotechnical aspects of telecommunication equipment
CLC/TC 216	Gas detectors
CLC/TC 21X	Secondary cells and batteries
CLC/TC 22X	Power electronics
CLC/TC 23BX	Switches, boxes and enclosures for household and similar purposes, plugs and socket outlet for D.C.
CLC/TC 23E	Circuit breakers and similar devices for household and similar applications
CLC/TC 23H	Plugs, Socket-outlets and Couplers for industrial and similar applications, and for Electric Vehicles
CLC/TC 26	Electric welding
CLC/TC 31	Electrical apparatus for potentially explosive atmospheres
CLC/TC 34	Lamps and related equipment
CLC/TC 36A	Insulated bushings
CLC/TC 37A	Low voltage surge protective devices
CLC/TC 38	Instrument transformers
CLC/TC 40XA	Capacitors and EMI suppression components
CLC/TC 40XB	Resistors
CLC/TC 44X	Safety of machinery: electrotechnical aspects
CLC/TC 45AX	Instrumentation, control and electrical power systems of nuclear facilities
CLC/TC 45B	Radiation protection instrumentation
CLC/TC 46X	Communication cables
CLC/TC 55	Winding wires
CLC/TC 57	Power systems management and associated information exchange

(continued)

Table 7.5 (continued)

Committee	Title
CLC/TC 59X	Performance of household and similar electrical appliances
CLC/TC 61	Safety of household and similar electrical appliances
CLC/TC 62	Electrical equipment in medical practice
CLC/TC 64	Electrical installations and protection against electric shock
CLC/TC 65X	Industrial-process measurement, control and automation
CLC/TC 66X	Safety of measuring, control, and laboratory equipment
CLC/TC 69X	Electrical systems for electric road vehicles
CLC/TC 72	Automatic electrical controls
CLC/TC 76	Optical radiation safety and laser equipment
CLC/TC 78	Equipment and tools for live working
CLC/TC 79	Alarm systems
CLC/TC 7X	Overhead electrical conductors
CLC/TC 81X	Lightning protection
CLC/TC 82	Solar photovoltaic energy systems
CLC/TC 85X	Measuring equipment for electrical and electromagnetic quantities
CLC/TC 86A	Optical fibres and optical fibre cables
CLC/TC 86BXA	Fibre optic interconnect, passive and connectorised components
CLC/TC 88	Wind turbines
CLC/TC 8X	System aspects of electrical energy supply
CLC/TC 94	Relays
CLC/TC 95X	Measuring relays and protection equipment
CLC/TC 99X	Power installations exceeding 1 kV a.c. (1.5 kV d.c.)
CLC/TC 9X	Electrical and electronic applications for railways

Table 7.6 List of CENELEC sub-committees

Committee	Title
CLC/SC 18XC	Subsea equipment
CLC/SC 205A	Mains communicating systems
CLC/SC 31-1	Installation rules
CLC/SC 31-2	Flameproof enclosures “d”
CLC/SC 31-3	Intrinsically safe apparatus and systems “i”
CLC/SC 31-4	Increased safety “e”
CLC/SC 31-5	Apparatus type of protection “n”
CLC/SC 31-7	Pressurization and other techniques
CLC/SC 31-8	Electrostatic painting and finishing equipment
CLC/SC 31-9	Electrical apparatus for the detection and measurement of combustible gases to be used in industrial and commercial potentially explosive atmospheres
CLC/SC 46XA	Coaxial cables
CLC/SC 46XC	Multicore, multipair and quad data communication cables
CLC/SC 9XA	Communication, signalling and processing systems
CLC/SC 9XB	Electrical, electronic and electromechanical material on board rolling stock, including associated software
CLC/SC 9XC	Electric supply and earthing systems for public transport equipment and ancillary apparatus (fixed installations)

Table 7.7 The list of CENELEC BTTF

Committee	Title
CLC/BTTF 116-2	Alcohol interlocks
CLC/BTTF 128-2	Erection and operation of electrical test equipment
CLC/BTTF 129-1	Thermal resistant aluminium alloy wire for overhead line conductor
CLC/BTTF 132-1	Aluminium conductors steel supported (ACSS type) for overhead electrical lines
CLC/BTTF 132-2	Revision of EN 50156 "Electrical equipment for furnaces and ancillary equipment"
CLC/BTTF 133-1	Sound systems for emergency purposes which are not part of fire detection and alarm systems
CLC/BTTF 146-1	Losses of small transformers: methods of measurement, marking and other requirements related to eco-design regulation
CLC/BTTF 157-1	Public address and general emergency alarm systems
CLC/BTTF 160-1	Recurrent Test of Electrical Equipment
CLC/BTTF 60-1	Assembly of electronic equipment
CLC/BTTF 62-3	Operation of electrical installations
CLC/BTTF 69-3	Road traffic signal systems

Table 7.8 List of CENELEC BTWG

Committee	Title
CLC/BTWG 126-2	Internal regulations
CLC/BTWG 128-3	BT efficiency
CLC/BTWG 138-1	CENELEC/BT action plan
CLC/BTWG 143-1	LVD standardization in the EU regulatory framework
CLC/BTWG 154-1	EMC OJEU listing

All CENELEC members obliged themselves for EN use and final issue in all country members. In addition to EN and HD, CENELEC publishes CENELEC Workshop Agreement, CWA, to shorten the development time for the EN or the HD.

CENELEC deliverables are both the voluntary and harmonized standards. In the year 2019, CENELEC had a total figure of 7305 active standards, whereas in 2016, 2017 and 2018, the numbers were 6857, 7026 and 7085, respectively. During the year 2019, 463 standards, 20 technical reports, 5 Technical Specifications, 1 guide, and 2 CENELEC workshop agreements, in total 491 deliverables, were published. The number of total active standards in 2017 (Column B, blue colour), 2018 (Column C, red colour) and 2019 (Column D, yellow colour), shown in Fig. 7.9, does not increase linearly with the published standards per year, because a certain number of withdrawn standards are then deleted from the list (i.e., they become non-active standards). Indeed, this is visible in the net increase at year end.

Figure 7.10 shows the total number of active CENELEC deliverables at the end of 2019. The ENVs/Reports/ESs/CECC Specifications are not published any more, which means that their number stays fixed, unless withdrawn.

The level of equivalence between CENELEC and IEC standards published in 2019 is shown in Fig. 7.11. (Identical to the IEC, 373; Based on the IEC, 13; Purely European, 104).

The level of equivalence between IEC and CENELEC for the overall deliverables at the end of 2019 is shown in Fig. 7.12 (Identical to the IEC, 4570; Based on the IEC, 405; Purely European, 1246).

Figure 7.13 shows the number of harmonized deliverables at the end of 2018 (1510) in comparison to total CENELEC deliverables (7352).

As an example, Table 7.9 provides a list of harmonized standards giving presumption of conformity for "Electromagnetic Compatibility" Directive, Directive 2014/30/EU [52] of the European Parliament and of the Council of February 26, 2014, on the harmonization of the laws of the Member States relating to electromagnetic compatibility (recast) that is applicable from April 20, 2016 (OJ L 96, March 29, 2014).

CENELEC European partners are given in Table 7.10.

International and supranational partners of CENELEC are:

- CEN, in Brussels, Belgium
- ETSI, in Sophia Antipolis, France
- IEC, in Geneva, Switzerland
- ISO, in Geneva, Switzerland
- ITU, in Geneva, Switzerland
- ABNT, Associação Brasileira de Normas Técnicas, in Sao Paolo, Brazil [53]
- ANSI, American National Standards Institute, in Washington, DC, USA [54]
- CANENA (span. Consejo de Armonizacion de Normas Electrotecnicas de las Naciones en las Americas, eng. Council for Harmonization of Electrotechnical Standards of the Nations in the Americas), in Rosslyn, USA [55]
- JISC (eng. Japanese Industrial Standards Committee), in Tokyo, Japan [56]

7.2.3 CEN-CENELEC Joint Activities

CEN and CENELEC are both ESOs. Nowadays, technical activities are not only in one field, because of the growing interdisciplinary products that surround us. Therefore, CEN and CENELEC have set objectives to be reached by 2020, namely: global influence, supranational relevance, wider recognition, network of excellence, innovation and growth, and a sustainable standardization system.

Both the CEN and CENELEC are independent in governing respective technical areas: General Assembly, Administrative Board, Technical Board, Advisory Bodies, and Technical Bodies (Technical Committees, Sub-Committees, and Working Groups). However, in order to facilitate cooperation on strategic matters in the developing strategic areas of strong interdisciplinarity and transdisciplinarity, such as an example of a smart city, CEN and CENELEC decided to share a joint structure of the CEN-CENELEC Presidential Committee (CCPC). CCPC consists of the two Presidents of CEN and CENELEC, the Presidents-Elect, all the Vice-Presidents, and the Director General of CEN and CENELEC.

The point of the central coordination, daily activities, and the promotion of all the CEN and CENELEC activities is also the joint CEN-CENELEC Management Centre,

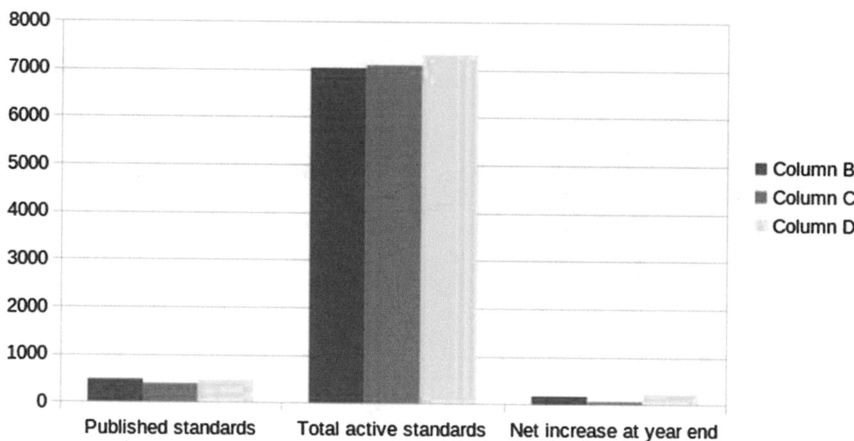

Fig. 7.9 CENELEC active and published standards and the net increase for 2017, 2018 and 2019 in December 2019

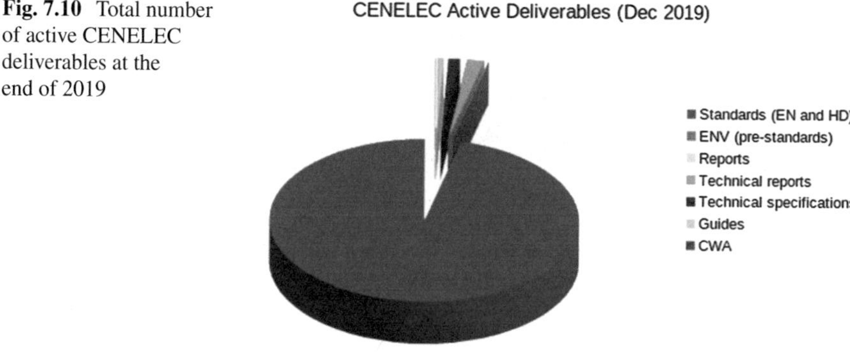

Fig. 7.10 Total number of active CENELEC deliverables at the end of 2019

Fig. 7.11 The level of equivalence between CENELEC and IEC standards published in 2019

Fig. 7.12 Level of equivalence between CENELEC and IEC deliverables for the overall deliverables at the end of 2019

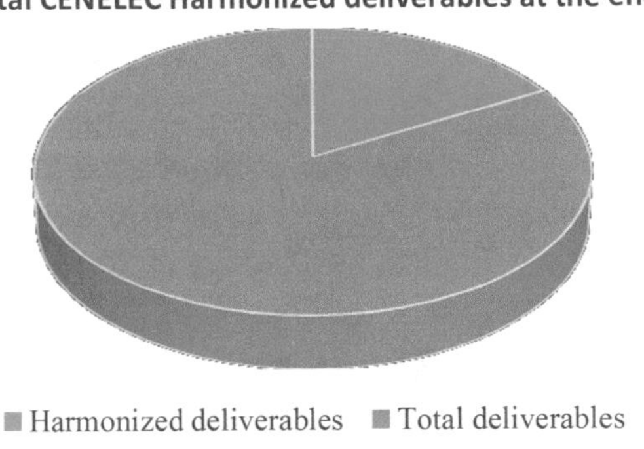

Fig. 7.13 Number of harmonized deliverables at the end of 2018 in comparison to total CENELEC deliverables

CCMC. CCMC corresponds with the European Commission and the EFTA Secretariat. It handles the tasks assigned by the both CEN and CENELEC General Assemblies, as well as by the CEN and CENELEC Technical and Administrative Boards.

The number of joint activities of CEN and CENELEC is growing, but for the moment the CEN-CENELEC sectors are accessibility, air and space, climate change adaptation, defense, security and privacy, energy management and energy efficiency, health and safety, healthcare, ICT, machinery, measuring instruments, smart living, and transport.

For example, the task of the CEN-CENELEC sector on energy management and energy efficiency is to help businesses and consumers to use energy more efficiently by supporting energy users with the necessary tools to analyze and adapt their energy consumption patterns. The standardization activities are in the areas of eco-design, energy labelling, energy management, smart grids, and smart meters.

CEN and CENELEC are working together in many coordination and focus groups. A very important group is a series of CEN-CENELEC (CEN/CLC) Guides. The CEN/CLC/Guide is a reference document that gives orientation, advice, or

Table 7.9 List of CENELEC harmonized standards giving presumption of conformity for "Electromagnetic Compatibility" Directive

Reference and title of the standard(and reference document)	First publication OJ
EN 50065-1:2011 Signalling on low-voltage electrical installations in the frequency range 3 kHz to 148,5 kHz - Part 1: General requirements, frequency bands and electromagnetic disturbances	13/05/2016
EN 50065-2-1:2003 Signalling on low-voltage electrical installations in the frequency range 3 kHz to 148,5 kHz - Part 2-1: Immunity requirements for mains communications equipment and systems operating in the range of frequencies 95 kHz to 148,5 kHz and intended for use in residential, commercial and light industrial environments	13/05/2016
EN 50065-2-1:2003/AC:2003	13/05/2016
EN 50065-2-1:2003/A1:2005	13/05/2016
EN 50065-2-2:2003 Signalling on low-voltage electrical installations in the frequency range 3 kHz to 148,5 kHz - Part 2-2: Immunity requirements for mains communications equipment and systems operating in the range of frequencies 95 kHz to 148,5 kHz and intended for use in industrial environments	13/05/2016
EN 50065-2-2:2003/A1:2005	13/05/2016
EN 50065-2-2:2003/AC:2003	13/05/2016
EN 50065-2-2:2003/A1:2005/AC:2006	13/05/2016
EN 50065-2-3:2003 Signalling on low-voltage electrical installations in the frequency range 3 kHz to 148,5 kHz - Part 2-3: Immunity requirements for mains communications equipment and systems operating in the range of frequencies 3 kHz to 95 kHz and intended for use by electricity suppliers and distributors	13/05/2016
EN 50065-2-3:2003/A1:2005	13/05/2016
EN 50065-2-3:2003/AC:2003	13/05/2016
EN 50083-2:2012 Cable networks for television signals, sound signals and interactive services - Part 2: Electromagnetic compatibility for equipment	13/05/2016
EN 50083-2:2012/A1:2015	13/05/2016
EN 50121-1:2006 Railway applications - Electromagnetic compatibility - Part 1: General	12/08/2016
EN 50121-1:2006/AC:2008	12/08/2016
EN 50121-2:2006 Railway applications - Electromagnetic compatibility - Part 2: Emission of the whole railway system to the outside world	12/08/2016
EN 50121-2:2006/AC:2008	12/08/2016
EN 50121-3-1:2017 (new) Railway applications - Electromagnetic compatibility - Part 3-1: Rolling stock - Train and complete vehicle	The first publication
EN 50121-3-2:2016 (new) Railway applications - Electromagnetic compatibility - Part 3-2: Rolling stock - Apparatus	The first publication

(continued)

Table 7.9 (continued)

Reference and title of the standard(and reference document)	First publication OJ
EN 50121-4:2016 (new) Railway applications - Electromagnetic compatibility - Part 4: Emission and immunity of the signalling and telecommunications apparatus	The first publication
EN 50121-5:2017 (new) Railway applications - Electromagnetic compatibility - Part 5: Emission and immunity of fixed power supply installations and apparatus	The first publication
EN 50130-4:2011 Alarm systems - Part 4: Electromagnetic compatibility - Product family standard: Immunity requirements for components of fire, intruder, hold up, CCTV, access control and social alarm systems	13/05/2016
EN 50148:1995 Electronic taximeters	13/05/2016
EN 50270:2015 (new) Electromagnetic compatibility - Electrical apparatus for the detection and measurement of combustible gases, toxic gases or oxygen	The first publication
EN 50270:2015/AC:2016-08 (new)	The first publication
EN 50293:2012 Road traffic signal systems - Electromagnetic compatibility	13/05/2016
EN 50370-1:2005 Electromagnetic compatibility (EMC) - Product family standard for machine tools - Part 1: Emission	13/05/2016
EN 50370-2:2003 Electromagnetic compatibility (EMC) - Product family standard for machine tools - Part 2: Immunity	13/05/2016
EN 50412-2-1:2005 Power line communication apparatus and systems used in low-voltage installations in the frequency range 1,6 MHz to 30 MHz - Part 2-1: Residential, commercial and industrial environment - Immunity requirements	13/05/2016
EN 50412-2-1:2005/AC:2009	13/05/2016
EN 50428:2005 Switches for household and similar fixed electrical installations - Collateral standard - Switches and related accessories for use in home and building electronic systems (HBES)	13/05/2016
EN 50428:2005/A1:2007	13/05/2016
EN 50428:2005/A2:2009	13/05/2016
EN 50470-1:2006 Electricity metering equipment (a.c.) - Part 1: General requirements, tests and test conditions - Metering equipment (class indexes A, B and C)	13/05/2016
EN 50490:2008 Electrical installations for lighting and beaconing of aerodromes - Technical requirements for aeronautical ground lighting control and monitoring systems - Units for selective switching and monitoring of individual lamps	13/05/2016

(continued)

Table 7.9 (continued)

Reference and title of the standard(and reference document)	First publication OJ
EN 50491-5-1:2010 General requirements for Home and Building Electronic Systems (HBES) and Building Automation and Control Systems (BACS) - Part 5-1: EMC requirements, conditions and test set-up	13/05/2016
EN 50491-5-2:2010 General requirements for Home and Building Electronic Systems (HBES) and Building Automation and Control Systems (BACS) - Part 5-2: EMC requirements for HBES/BACS used in residential, commercial and light industry environment	13/05/2016
EN 50491-5-3:2010 General requirements for Home and Building Electronic Systems (HBES) and Building Automation and Control Systems (BACS) - Part 5-3: EMC requirements for HBES/BACS used in industry environment	13/05/2016
EN 50498:2010 Electromagnetic compatibility (EMC) - Product family standard for aftermarket electronic equipment in vehicles	13/05/2016
EN 50512:2009 Electrical installations for lighting and beaconing of aerodromes - Advanced Visual Docking Guidance Systems (A-VDGS)	13/05/2016
EN 50529-1:2010 EMC Network Standard - Part 1: Wire-line telecommunications networks using telephone wires	13/05/2016
EN 50529-2:2010 EMC Network Standard - Part 2: Wire-line telecommunications networks using coaxial cables	13/05/2016
EN 50550:2011 Power frequency overvoltage protective device for household and similar applications (POP)	13/05/2016
EN 50550:2011/AC:2012	13/05/2016
EN 50550:2011/A1:2014	13/05/2016
EN 50557:2011 Requirements for automatic reclosing devices (ARDs) for circuit breakers-RCBOs-RCCBs for household and similar uses	13/05/2016
EN 50561-1:2013 Power line communication apparatus used in low-voltage installations - Radio disturbance characteristics - Limits and methods of measurement - Part 1: Apparatus for in-home use	13/05/2016
EN 55011:2009 Industrial, scientific and medical equipment - Radio-frequency disturbance characteristics - Limits and methods of measurementCISPR 11:2009 (Modified)	13/05/2016
EN 55011:2009/A1:2010CISPR 11:2009/A1:2010	13/05/2016
EN 55012:2007 Vehicles, boats and internal combustion engines - Radio disturbance characteristics - Limits and methods of measurement for the protection of off-board receiversCISPR 12:2007	13/05/2016
EN 55012:2007/A1:2009CISPR 12:2007/A1:2009	13/05/2016

(continued)

Table 7.9 (continued)

Reference and title of the standard(and reference document)	First publication OJ
EN 55014-1:2006 Electromagnetic compatibility - Requirements for household appliances, electric tools and similar apparatus - Part 1: EmissionCISPR 14-1:2005	13/05/2016
EN 55014-1:2006/A2:2011CISPR 14-1:2005/A2:2011	13/05/2016
EN 55014-1:2006/A1:2009CISPR 14-1:2005/A1:2008	13/05/2016
EN 55014-2:1997 Electromagnetic compatibility - Requirements for household appliances, electric tools and similar apparatus - Part 2: Immunity - Product family standardCISPR 14-2:1997	12/08/2016
EN 55014-2:1997/A2:2008CISPR 14-2:1997/A2:2008	12/08/2016
EN 55014-2:1997/AC:1997	12/08/2016
EN 55014-2:1997/A1:2001CISPR 14-2:1997/A1:2001	12/08/2016
EN 55015:2013 Limits and methods of measurement of radio disturbance characteristics of electrical lighting and similar equipmentCISPR 15:2013 + IS1:2013 + IS2:2013	13/05/2016
EN 55024:2010 Information technology equipment - Immunity characteristics - Limits and methods of measurementCISPR 24:2010	13/05/2016
EN 55032:2012 Electromagnetic compatibility of multimedia equipment - Emission requirementsCISPR 32:2012	12/08/2016
EN 55032:2012/AC:2013	12/08/2016
EN 55103-1:2009 Electromagnetic compatibility - Product family standard for audio, video, audio-visual and entertainment lighting control apparatus for professional use - Part 1: Emissions	12/08/2016
EN 55103-1:2009/A1:2012	12/08/2016
EN 55103-2:2009 Electromagnetic compatibility - Product family standard for audio, video, audio-visual and entertainment lighting control apparatus for professional use - Part 2: Immunity	13/05/2016
EN 60034-1:2010 Rotating electrical machines - Part 1: Rating and performanceIEC 60034-1:2010 (Modified)	13/05/2016
EN 60034-1:2010/AC:2010	13/05/2016
EN 60204-31:2013 Safety of machinery - Electrical equipment of machines - Part 31: Particular safety and EMC requirements for sewing machines, units and systemsIEC 60204-31:2013	13/05/2016
EN 60255-26:2013 Measuring relays and protection equipment - Part 26: Electromagnetic compatibility requirementsIEC 60255-26:2013	13/05/2016
EN 60255-26:2013/AC:2013	13/05/2016

(continued)

Table 7.9 (continued)

Reference and title of the standard(and reference document)	First publication OJ
EN 60669-2-1:2004 Switches for household and similar fixed electrical installations - Part 2-1: Particular requirements - Electronic switchesIEC 60669-2-1:2002 (Modified) + IS1:2011 + IS2:2012	13/05/2016
EN 60669-2-1:2004/AC:2007	13/05/2016
EN 60669-2-1:2004/A1:2009IEC 60669-2-1:2002/A1:2008 (Modified)	13/05/2016
EN 60669-2-1:2004/A12:2010	13/05/2016
EN 60730-1:2011 Automatic electrical controls for household and similar use - Part 1: General requirementsIEC 60730-1:2010 (Modified)	13/05/2016
EN 60730-2-5:2002 Automatic electrical controls for household and similar use - Part 2-5: Particular requirements for automatic electrical burner control systemsIEC 60730-2-5:2000 (Modified)	12/08/2016
EN 60730-2-5:2002/A11:2005	12/08/2016
EN 60730-2-5:2002/A1:2004IEC 60730-2-5:2000/A1:2004 (Modified)	12/08/2016
EN 60730-2-5:2002/A2:2010IEC 60730-2-5:2000/A2:2008 (Modified)	12/08/2016
EN 60730-2-6:2008 Automatic electrical controls for household and similar use - Part 2-6: Particular requirements for automatic electrical pressure sensing controls including mechanical requirementsIEC 60730-2-6:2007 (Modified)	12/08/2016
EN 60730-2-7:2010 Automatic electrical controls for household and similar use - Part 2-7: Particular requirements for timers and time switchesIEC 60730-2-7:2008 (Modified)	13/05/2016
EN 60730-2-7:2010/AC:2011	13/05/2016
EN 60730-2-8:2002 Automatic electrical controls for household and similar use - Part 2-8: Particular requirements for electrically operated water valves, including mechanical requirementsIEC 60730-2-8:2000 (Modified)	13/05/2016
EN 60730-2-8:2002/A1:2003IEC 60730-2-8:2000/A1:2002 (Modified)	13/05/2016
EN 60730-2-9:2010 Automatic electrical controls for household and similar use - Part 2-9: Particular requirements for temperature sensing controlsIEC 60730-2-9:2008 (Modified)	13/05/2016
EN 60730-2-14:1997 Automatic electrical controls for household and similar use - Part 2-14: Particular requirements for electric actuatorsIEC 60730-2-14:1995 (Modified)	13/05/2016
EN 60730-2-14:1997/A1:2001IEC 60730-2-14:1995/A1:2001	13/05/2016
EN 60730-2-15:2010 Automatic electrical controls for household and similar use - Part 2-15: Particular requirements for automatic electrical air flow, water flow and water level sensing controlsIEC 60730-2-15:2008 (Modified)	13/05/2016
EN 60870-2-1:1996 Telecontrol equipment and systems - Part 2: Operating conditions - Section 1: Power supply and electromagnetic compatibilityIEC 60870-2-1:1995	13/05/2016

(continued)

Table 7.9 (continued)

Reference and title of the standard(and reference document)	First publication OJ
EN 60945:2002 Maritime navigation and radiocommunication equipment and systems - General requirements - Methods of testing and required test resultsIEC 60945:2002	13/05/2016
EN 60947-1:2007 Low-voltage switchgear and controlgear - Part 1: General rulesIEC 60947-1:2007	13/05/2016
EN 60947-1:2007/A2:2014IEC 60947-1:2007/A2:2014	13/05/2016
EN 60947-1:2007/A1:2011IEC 60947-1:2007/A1:2010	13/05/2016
EN 60947-2:2006 Low-voltage switchgear and controlgear - Part 2: Circuit-breakersIEC 60947-2:2006	13/05/2016
EN 60947-2:2006/A1:2009IEC 60947-2:2006/A1:2009	13/05/2016
EN 60947-2:2006/A2:2013IEC 60947-2:2006/A2:2013	13/05/2016
EN 60947-3:2009 Low-voltage switchgear and controlgear - Part 3: Switches, disconnectors, switch-disconnectors and fuse-combination unitsIEC 60947-3:2008	13/05/2016
EN 60947-3:2009/A1:2012IEC 60947-3:2008/A1:2012	13/05/2016
EN 60947-4-1:2010 Low-voltage switchgear and controlgear - Part 4-1: Contactors and motor-starters - Electromechanical contactors and motor-startersIEC 60947-4-1:2009	13/05/2016
EN 60947-4-1:2010/A1:2012IEC 60947-4-1:2009/A1:2012	13/05/2016
EN 60947-4-2:2012 Low-voltage switchgear and controlgear - Part 4-2: Contactors and motor-starters - AC semiconductor motor controllers and startersIEC 60947-4-2:2011	13/05/2016
EN 60947-4-3:2014 Low-voltage switchgear and controlgear - Part 4-3: Contactors and motor-starters - AC semiconductor controllers and contactors for non-motor loadsIEC 60947-4-3:2014	13/05/2016
EN 60947-5-1:2004 Low-voltage switchgear and controlgear - Part 5-1: Control circuit devices and switching elements - Electromechanical control circuit devicesIEC 60947-5-1:2003	13/05/2016
EN 60947-5-1:2004/AC:2005	13/05/2016
EN 60947-5-1:2004/A1:2009IEC 60947-5-1:2003/A1:2009	13/05/2016
EN 60947-5-1:2004/AC:2004	13/05/2016
EN 60947-5-2:2007 Low-voltage switchgear and controlgear - Part 5-2: Control circuit devices and switching elements - Proximity switchesIEC 60947-5-2:2007	13/05/2016
EN 60947-5-2:2007/A1:2012IEC 60947-5-2:2007/A1:2012	13/05/2016
EN 60947-5-3:1999 Low-voltage switchgear and controlgear - Part 5-3: Control circuit devices and switching elements - Requirements for proximity devices with defined behaviour under fault conditions (PDF)IEC 60947-5-3:1999	13/05/2016
EN 60947-5-3:1999/A1:2005IEC 60947-5-3:1999/A1:2005	13/05/2016

(continued)

Table 7.9 (continued)

Reference and title of the standard(and reference document)	First publication OJ
EN 60947-5-6:2000 Low-voltage switchgear and controlgear - Part 5-6: Control circuit devices and switching elements - DC interface for proximity sensors and switching amplifiers (NAMUR)IEC 60947-5-6:1999	13/05/2016
EN 60947-5-7:2003 Low-voltage switchgear and controlgear - Part 5-7: Control circuit devices and switching elements - Requirements for proximity devices with analogue outputIEC 60947-5-7:2003	13/05/2016
EN 60947-5-9:2007 Low-voltage switchgear and controlgear - Part 5-9: Control circuit devices and switching elements - Flow rate switchesIEC 60947-5-9:2006	13/05/2016
EN 60947-6-1:2005 Low-voltage switchgear and controlgear - Part 6-1: Multiple function equipment - Transfer switching equipmentIEC 60947-6-1:2005	13/05/2016
EN 60947-6-1:2005/A1:2014IEC 60947-6-1:2005/A1:2013	13/05/2016
EN 60947-6-2:2003 Low-voltage switchgear and controlgear - Part 6-2: Multiple function equipment - Control and protective switching devices (or equipment) (CPS)IEC 60947-6-2:2002	13/05/2016
EN 60947-6-2:2003/A1:2007IEC 60947-6-2:2002/A1:2007	13/05/2016
EN 60947-8:2003 Low-voltage switchgear and controlgear - Part 8: Control units for built-in thermal protection (PTC) for rotating electrical machinesIEC 60947-8:2003	13/05/2016
EN 60947-8:2003/A2:2012IEC 60947-8:2003/A2:2011	13/05/2016
EN 60947-8:2003/A1:2006IEC 60947-8:2003/A1:2006	13/05/2016
EN 60974-10:2014 Arc welding equipment - Part 10: Electromagnetic compatibility (EMC) requirementsIEC 60974-10:2014	13/05/2016
EN 61000-3-2:2014 Electromagnetic compatibility (EMC) - Part 3-2: Limits - Limits for harmonic current emissions (equipment input current ≤16 A per phase)IEC 61000-3-2:2014	13/05/2016
EN 61000-3-3:2013 Electromagnetic compatibility (EMC) - Part 3-3: Limits - Limitation of voltage changes, voltage fluctuations and flicker in public low-voltage supply systems, for equipment with rated current <= 16 A per phase and not subject to conditional connectionIEC 61000-3-3:2013	13/05/2016
EN 61000-3-11:2000 Electromagnetic compatibility (EMC) - Part 3-11: Limits - Limitation of voltage changes, voltage fluctuations and flicker in public low-voltage supply systems - Equipment with rated current <= 75 A and subject to conditional connectionIEC 61000-3-11:2000	13/05/2016
EN 61000-3-12:2011 Electromagnetic compatibility (EMC) - Part 3-12: Limits - Limits for harmonic currents produced by equipment connected to public low-voltage systems with input current >16 A and <= 75 A per phaseIEC 61000-3-12:2011 + IS1:2012	13/05/2016

(continued)

Table 7.9 (continued)

Reference and title of the standard(and reference document)	First publication OJ
EN 61000-6-1:2007 Electromagnetic compatibility (EMC) - Part 6-1: Generic standards - Immunity for residential, commercial and light-industrial environmentsIEC 61000-6-1:2005	13/05/2016
EN 61000-6-2:2005 Electromagnetic compatibility (EMC) - Part 6-2: Generic standards - Immunity for industrial environmentsIEC 61000-6-2:2005	13/05/2016
EN 61000-6-2:2005/AC:2005	13/05/2016
EN 61000-6-3:2007 Electromagnetic compatibility (EMC) - Part 6-3: Generic standards - Emission standard for residential, commercial and light-industrial environmentsIEC 61000-6-3:2006	13/05/2016
EN 61000-6-3:2007/A1:2011IEC 61000-6-3:2006/A1:2010	13/05/2016
EN 61000-6-3:2007/A1:2011/AC:2012	13/05/2016
EN 61000-6-4:2007 Electromagnetic compatibility (EMC) - Part 6-4: Generic standards - Emission standard for industrial environmentsIEC 61000-6-4:2006	13/05/2016
EN 61000-6-4:2007/A1:2011IEC 61000-6-4:2006/A1:2010	13/05/2016
EN 61000-6-5:2015 (new) Electromagnetic compatibility (EMC) - Part 6-5: Generic standards - Immunity for equipment used in power station and substation environmentIEC 61000-6-5:2015	The first publication
EN 61008-1:2012 Residual current operated circuit-breakers without integral overcurrent protection for household and similar uses (RCCBs) - Part 1: General rulesIEC 61008-1:2010 (Modified)	13/05/2016
EN 61008-1:2012/A1:2014IEC 61008-1:2010/A1:2012 (Modified)	13/05/2016
EN 61009-1:2012 Residual current operated circuit-breakers with integral overcurrent protection for household and similar uses (RCBOs) - Part 1: General rulesIEC 61009-1:2010 (Modified)	13/05/2016
EN 61131-2:2007 Programmable controllers - Part 2: Equipment requirements and testsIEC 61131-2:2007	13/05/2016
EN 61204-3:2000 Low voltage power supplies, d.c. output - Part 3: Electromagnetic compatibility (EMC)IEC 61204-3:2000	13/05/2016
EN 61326-1:2013 Electrical equipment for measurement, control and laboratory use - EMC requirements - Part 1: General requirementsIEC 61326-1:2012	13/05/2016
EN 61326-2-1:2013 Electrical equipment for measurement, control and laboratory use - EMC requirements - Part 2-1: Particular requirements - Test configurations, operational conditions and performance criteria for sensitive test and measurement equipment for EMC unprotected applicationsIEC 61326-2-1:2012	13/05/2016

(continued)

Table 7.9 (continued)

Reference and title of the standard(and reference document)	First publication OJ
EN 61326-2-2:2013 Electrical equipment for measurement, control and laboratory use - EMC requirements - Part 2-2: Particular requirements - Test configurations, operational conditions and performance criteria for portable test, measuring and monitoring equipment used in low-voltage distribution systemsIEC 61326-2-2:2012	13/05/2016
EN 61326-2-3:2013 Electrical equipment for measurement, control and laboratory use - EMC requirements - Part 2-3: Particular requirements - Test configuration, operational conditions and performance criteria for transducers with integrated or remote signal conditioningIEC 61326-2-3:2012	13/05/2016
EN 61326-2-4:2013 Electrical equipment for measurement, control and laboratory use - EMC requirements - Part 2-4: Particular requirements - Test configurations, operational conditions and performance criteria for insulation monitoring devices according to IEC 61557-8 and for equipment for insulation fault location according to IEC 61557-9IEC 61326-2-4:2012	13/05/2016
EN 61326-2-5:2013 Electrical equipment for measurement, control and laboratory use - EMC requirements - Part 2-5: Particular requirements - Test configurations, operational conditions and performance criteria for devices with field bus interfaces according to IEC 61784-1IEC 61326-2-5:2012	13/05/2016
EN 61439-1:2011 Low-voltage switchgear and controlgear assemblies - Part 1: General rulesIEC 61439-1:2011	13/05/2016
EN 61439-2:2011 Low-voltage switchgear and controlgear assemblies - Part 2: Power switchgear and controlgear assembliesIEC 61439-2:2011	13/05/2016
EN 61439-3:2012 Low-voltage switchgear and controlgear assemblies - Part 3: Distribution boards intended to be operated by ordinary persons (DBO)IEC 61439-3:2012	13/05/2016
EN 61439-4:2013 Low-voltage switchgear and controlgear assemblies - Part 4: Particular requirements for assemblies for construction sites (ACS)IEC 61439-4:2012	13/05/2016
EN 61439-5:2011 Low-voltage switchgear and controlgear assemblies - Part 5: Assemblies for power distribution in public networksIEC 61439-5:2010	12/08/2016
EN 61439-6:2012 Low-voltage switchgear and controlgear assemblies - Part 6: Busbar trunking systems (busways)IEC 61439-6:2012	13/05/2016
EN 61543:1995 Residual current-operated protective devices (RCDs) for household and similar use - Electromagnetic compatibilityIEC 61543:1995	13/05/2016
EN 61543:1995/A2:2006IEC 61543:1995/A2:2005	13/05/2016
EN 61543:1995/A11:2003/AC:2004	13/05/2016
EN 61543:1995/AC:1997	13/05/2016

(continued)

Table 7.9 (continued)

Reference and title of the standard(and reference document)	First publication OJ
EN 61543:1995/A11:2003	13/05/2016
EN 61543:1995/A12:2005	13/05/2016
EN 61547:2009 Equipment for general lighting purposes - EMC immunity requirementsIEC 61547:2009 + IS1:2013	13/05/2016
EN 61557-12:2008 Electrical safety in low voltage distribution systems up to 1 000 V a.c. and 1 500 V d.c. - Equipment for testing, measuring or monitoring of protective measures - Part 12: Performance measuring and monitoring devices (PMD)IEC 61557-12:2007	13/05/2016
EN 61800-3:2004 Adjustable speed electrical power drive systems - Part 3: EMC requirements and specific test methodsIEC 61800-3:2004	13/05/2016
EN 61800-3:2004/A1:2012IEC 61800-3:2004/A1:2011	13/05/2016
EN 61812-1:2011 Time relays for industrial and residential use - Part 1: Requirements and testsIEC 61812-1:2011	13/05/2016
EN 62020:1998 Electrical accessories - Residual current monitors for household and similar uses (RCMs)IEC 62020:1998	13/05/2016
EN 62020:1998/A1:2005IEC 62020:1998/A1:2003 (Modified)	13/05/2016
EN 62026-1:2007 Low-voltage switchgear and controlgear - Controller-device interfaces (CDIs) - Part 1: General rulesIEC 62026-1:2007	13/05/2016
EN 62026-2:2013 Low-voltage switchgear and controlgear - Controller-device interfaces (CDIs) - Part 2: Actuator sensor interface (AS-i)IEC 62026-2:2008 (Modified)	13/05/2016
EN 62026-3:2009 Low-voltage switchgear and controlgear - Controller-device interfaces (CDIs) - Part 3: DeviceNetIEC 62026-3:2008	12/08/2016
EN 62026-7:2013 Low-voltage switchgear and controlgear - Controller-device interfaces (CDIs) - Part 7: CompoNetIEC 62026-7:2010 (Modified)	13/05/2016
EN 62040-2:2006 Uninterruptible power systems (UPS) - Part 2: Electromagnetic compatibility (EMC) requirementsIEC 62IEC 62040-2:2005	13/05/2016
EN 62040-2:2006/AC:2006	13/05/2016
EN 62052-11:2003 Electricity metering equipment (AC) - General requirements, tests and test conditions - Part 11: Metering equipmentIEC 62052-11:2003	13/05/2016
EN 62052-21:2004 Electricity metering equipment (a.c.) - General requirements, tests and test conditions - Part 21: Tariff and load control equipmentIEC 62052-21:2004	13/05/2016

(continued)

Table 7.9 (continued)

Reference and title of the standard(and reference document)	First publication OJ
EN 62053-11:2003 Electricity metering equipment (a.c.) - Particular requirements - Part 11: Electromechanical meters for active energy (classes 0,5, 1 and 2)IEC 62053-11:2003	13/05/2016
EN 62053-21:2003 Electricity metering equipment (a.c.) - Particular requirements - Part 21: Static meters for active energy (classes 1 and 2)IEC 62053-21:2003	13/05/2016
EN 62053-22:2003 Electricity metering equipment (a.c.) - Particular requirements - Part 22: Static meters for active energy (classes 0,2 S and 0,5 S)IEC 62053-22:2003	13/05/2016
EN 62053-23:2003 Electricity metering equipment (a.c.) - Particular requirements - Part 23: Static meters for reactive energy (classes 2 and 3)IEC 62053-23:2003	13/05/2016
EN 62054-11:2004 Electricity metering (a.c.) - Tariff and load control - Part 11: Particular requirements for electronic ripple control receiversIEC 62054-11:2004	13/05/2016
EN 62054-21:2004 Electricity metering (a.c.) - Tariff and load control - Part 21: Particular requirements for time switchesIEC 62054-21:2004	13/05/2016
EN 62135-2:2008 Resistance welding equipment - Part 2: Electromagnetic compatibility (EMC) requirementsIEC 62135-2:2007	12/08/2016
EN 62310-2:2007 Static transfer systems (STS) - Part 2: Electromagnetic compatibility (EMC) requirementsIEC 62310-2:2006 (Modified)	13/05/2016
EN 62423:2012 Type F and type B residual current operated circuit-breakers with and without integral overcurrent protection for household and similar usesIEC 62423:2009 (Modified)	13/05/2016
EN 62586-1:2014 Power quality measurement in power supply systems - Part 1: Power quality instruments (PQI)IEC 62586-1:2013	13/05/2016
EN 62586-2:2014 Power quality measurement in power supply systems - Part 2: Functional tests and uncertainty requirementsIEC 62586-2:2013	13/05/2016
EN 62606:2013 General requirements for arc fault detection devicesIEC IEC 62606:2013 (Modified)	13/05/2016

recommendations on policy and guidance to standard developers and writers. For example, in 2018 CEN/CLC published the CEN/CLC/Guide 22: "Guide on the organisational structure and processes for the assessment of the membership criteria of CEN and CENELEC." There are four CEN/CLC coordination groups (on adaptation to climate change, on light, on eco-design, and on eMobility), four joint CEN/CLC Sector Fora (on energy management, on machinery safety, on personal

Table 7.10 List of CENELEC European partners

Organization	Acronym
The European Association for the Co-ordination of Consumer Representation in Standardisation	ANEC
Home Appliance Europe	APPLiA
Coordinating Committee for the Associations of Manufacturers of Switchgear and Controlgear	CAPIEL
European Committee of Electrical Installation Equipment Manufacturers	CECAPI
European Cable Communications Association	Cable Europe
European Environmental Citizens Organisation for Standardisation	ECOS
European Trade Union Institute	ETUI
European Confederation of Associations of Manufacturers of Insulated Wires and Cables	EUROPACABLE
KNX Association	KNX
Europe's Technology Industries	ORGALIM
Small Business Standards	SBS
European Association of the Electricity Transmission and Distribution Equipment and Services Industry	T&D Europe

protective equipment, and for healthcare standards), one CEN/CLC Focus Group on Nuclear Energy, and three other CEN/CLC groups (SME Working Group, Societal Stakeholders Group, and, once important but now disbanded, Joint Working Group Education about Standardization).

7.2.4 KEYMARK

The story of a voluntary European quality mark for products and services, KEYMARK, started by Recommendation of EU Council to establish a European standard conformity mark in 1992. The development of the European system of signs of CEN-CENELEC continued in the period 1993–1994, and in 1995 the KEYMARK was already introduced. The first KEYMARK was given in Germany for thermal insulation products for buildings in the year 2000. Introduction of solar KEYMARK followed in 2004. The previously introduced CENCER mark, developed and owned by CEN, was transferred to KEYMARK in 2012. In 2015, the DIN CERTCO took over the KEYMARK Management. Introduction of Heat Pump KEYMARK followed in 2016. In 2018 was started the development of Security KEYMARK.

The KEYMARK's slogan is "Tested and certified once, accepted everywhere!" and it shows clearly the significance of the mark that demonstrates compliance with European Standards. The European Standards Organizations CEN and CENELEC are owners of KEYMARK, and the Certification Bodies issue it. KEYMARK symbol can be found on the web page [57].

In 2018 there are 2272 valid KEYMARKs. There are 23 European Standards (EN) with KEYMARK certificates. In KEYMARK certification 35 countries collaborate with total of 36 empowered certification bodies. Most of the given KEYMARK license fees go to the solar KEYMARK (64%). The rest of the licenses is distributed almost equally between heat pump KEYMARK (12%), insulation KEYMARK (10%), and thermostatic radiator valves (10%). Only 4% goes to other KEYMARK.

7.2.5 ETSI

The European Telecommunications Standards Institute (ETSI) is, together with CEN and CENELEC, the third partner in European Standards Organization (ESO) trio. ETSI issues globally applied standards in the field of information and communication technologies, which includes fixed, mobile, radio, transmitter, and Internet technologies. ETSI is a nonprofit organization with more than 850 member organizations worldwide from over 60 countries and 5 continents. It unifies national administrations, network operators, manufacturers, service operators, scientific institutions, users, and consultants.

ETSI issues between 2000 and 2500 standards every year, including the standards in the key global technologies areas, as GSM, 3G, 4G, 5G, DECT, smart cards, etc. From its beginning in 1988, ETSI has issued more than 30,000 standards.

Cooperation within ETSI results in the successful ICT standards in the field of mobile, fixed, and radiocommunications and in other areas such as:

- Aeronautical applications
- e-Health
- GRID and Clouds
- Human factors
- Radio and TV transmitting
- Satellite communications
- Security
- Smart cards
- Smart transport
- Testing and protocols

ETSI is based on consensus and on the work of Technical Committees that create standards and specifications with ETSI General Assembly and ETSI Board.

ETSI Directives define legal status, responsibilities, and detailed working procedure. For example, these are Statute, Board Working Procedures, ETSI Drafting Rules, Financial Regulations, Guide on Intellectual Property Rights, Guidelines for antitrust compliance, Guidelines for the implementation of Rules of Procedure, Information Policy, Powers and Functions delegated to the Board, Rules of Procedure, Technical Working Procedures, and Terms of Reference of the Operational Coordination Group (OCG).

ETSI management structure is based on:

- General Assembly, as the highest decision-making authority
- ETSI Board, as the executive body of the General Assembly
- Secretariat that provides support to ETSI Members and various committees

The General Assembly (GA) determines ETSI policy and strategy, deals with membership issues, appoints members of the ETSI Board, appoints Director General and Finance Committee members, approves changes to the Statutes and Rules of Procedure, endorses external agreements, and agrees budgets.

ETSI Board oversees Work Programme, endorses appointment of Technical Committee chairmen, advises GA on budget and finance issues, and approves Terms of Reference (ToR) for Technical Committees (TCs), ETSI projects, and Coordination groups.

The Secretariat has its address at Sophia Antipolis, France. It comprises 120 employees, providing technical, administrative, and logistical support with Director General as the head. Secretariat supports committees and projects and processes, approves, and publishes standards drafted by the committees. It also hosts ETSI committees, organizes workshops and events, and manages financial aspects, IT services, legal, human resources, and communication services. Centre for Testing and Interoperability provides support and assistance to the Technical Committees.

The technical standardization organization consists of various technical groups:

- ETSI Technical Committee (TC)
- ETSI Project (EP)
- ETSI Partnership Project (EPP)
- Industry Specification Group (ISG)
- Special Committee (SC)
- Specialist Task Force (STF)

ETSI Technical Committee (TC) starts its new activity with the definition of the so-called Work Item (new specific standardization task) which results in either a standard or a report. ETSI Work Programme is the sum of all the TCs' work programs. Operational Coordination Group (OCG) coordinates TCs.

The following TCs are active, as of May 2019:

- Access, Terminals, Transmission and Multiplexing (ATTM)
- Broadband Radio Access Networks (BRAN)
- Core Network and Interoperability Testing (INT)
- Cyber Security (CYBER)
- Digital Enhanced Cordless Telecommunications (DECT)
- Environmental Engineering (EE)
- EMC and Radio Spectrum Matters (ERM)
- Electronic Signatures and Infrastructures (ESI)
- Human Factors (HF)
- Intelligent Transport Systems (ITS)
- Integrated Broadband Cable Telecommunication Networks (CABLE)
- Lawful Interception (LI)

- Methods for Testing and Specification (MTS)
- Mobile Standards Group (MSG)
- Network Technologies (NTECH)
- Rail Telecommunications (RT)
- Reconfigurable Radio Systems (RRS)
- Telecommunications Equipment Safety (SAFETY)
- Satellite Earth Stations and Systems (SES)
- Smart Body Area Network (SMARTBAN)
- Smart Card Platform (SCP)
- Smart Machine-to-Machine Communications (SMARTM2M)
- Speech and Multimedia Transmission Quality (STQ)
- Terrestrial Trunked Radio and Critical Communications Evolution (TCCE)
- Joint Technical Committee (JTC) of the European Broadcasting Union (EBU), the European Committee for Electrotechnical Standardization (CENELEC), and ETSI

ETSI Project (EP) is established in order to meet specific market needs. Its lifetime is closely related to the market requirements. For example, as of May 2019, the EP is EP eHealth.

ETSI Partnership Project (EPP) enables cooperation of ETSI with other organizations to get the standards. As of May 2019, two EPPs exist: Third Generation Partnership Project (3GPP) and one M2M.

The Industry Specification Group (ISG) is an alternative of industry forum, because it is quickly and easily set up, in order to respond quickly to industry needs. ISGs are specific in ETSI, because of their own voting rules, own work program, and own membership (can be, but does not have to be, ETSI). ISG drafts, produces, and approves Group Specifications (GSs) or Group Reports (GRs), published by ETSI. The following ISGs are present:

- Augmented Reality Framework (ARF)
- City Digital Profile (CDP)
- Cross Cutting Context Information Management (CIM)
- Experiential Networked Intelligence (ENI)
- European Common Information Sharing Environment Service and Data Model (CDM)
- IPV6 Integration (IP6)
- Multi-Access Edge Computing (MEC)
- Millimetre Wave Transmission (MWT)
- Network Functions Virtualisation (NFV)
- Next Generation Protocols (NGP)
- Operational Energy Efficiency for Users (OEU)
- Permissioned Distributed Ledger (PDL)
- Quantum Key Distribution for Users (QKD)
- Zero Touch Network and Service Management (ZSM)

Special Committee (SC) coordinates and gathers requirements, so it does not draft standards and specifications. The following SCs are active as of May 2019:

- SC Emergency Telecommunications (EMTEL)
- SC User Group

ETSI has one open source project:

- Open Source Mano (OSM)

The specialist Task Force (STF) performs specific technical work under one of ETSI TCs in order to reply to the urgent market needs.

Chairs of the technical bodies are elected by the technical bodies and are established by the ETSI Board. The chairs are responsible for the total management of the technical body, its working groups, and execution of its working program. The responsibility for a work item has the expert group with the chairperson, who can also be a Rapporteur.

As explained in Chap. 4, ETSI publishes European Standard (EN), ETSI Standard (ES), ETSI Technical Specification (ETSI TS), ETSI Technical Report (ETSI TR), ETSI Guide (EG), ETSI Special Report (ETSI SR), ETSI Group Specification (ETSI GS), and ETSI Group Report (ETSI GR).

Only a standardization request from the European Commission (EC) or European Free Trade Association (EFTA) can start the work on the European Standard (EN). The Technical Committee (TC) drafts EN. ETSI's European National Standards Organization approves it.

ETSI has a wide membership and among them a lot of National Standards Organizations (NSOs). NSOs have a key role in adoption of standards, because they are the one who are excellent mediators between SMEs and microenterprises and the European world of standards. They carry out a public enquiry, which submits the national position (or the vote) on the standard and which ensures a smooth transposition of ENs into national standards that include a withdrawal of any conflicting national standard.

ETSI covers a wide area comprising ten sectors: Better Living with ICT, Connecting Things, Content Delivery, Home and Office, Interoperability, Networks, Public Safety, Security, Transportation, and Wireless Systems.

ETSI covers 47 various technologies: 5G, Aeronautical, Augmented Reality Framework, Automotive Intelligent Transport Systems (ITS), Broadband Cable Access, Broadband Wireless Access, Broadcast, Cyber security, Digital Enhanced Cordless Telecommunications (DECT), Digital Signature, eHealth, Electromagnetic Compatibility, Energy Efficiency (EE), Environmental Aspects, Experiential Networked Intelligence, Fixed Radio Links, Fixed-line Access, Human Factors and Accessibility, Internet of Things (IoT), Lawful Interception (LI), Maritime, Medical Devices, Mobile and Private Mobile Radio, Mobile Communications, Multi-access Edge Computing, Network Functions Virtualisation, Next Generation Protocols, Programme making and special events, Public safety and emergency communications, Quality of Service, Quantum Key Distribution, Quantum-Safe Cryptography, Radio, Radio LAN, Rail Communications, Reconfigurable Radio, Safety, Satellite Services, Security, Security Algorithms, Smart Appliances, Smart Body Area Networks, Smart Cards, Smart Cities, Testing Languages, TETRA, and Zero Touch Network and Service Management.

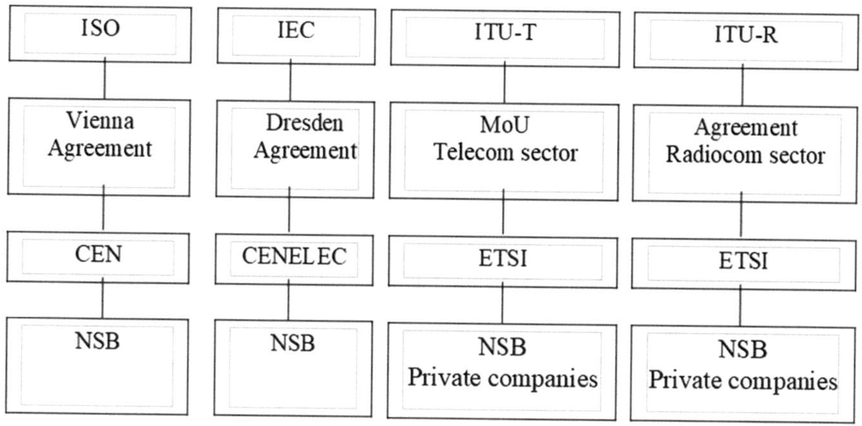

Fig. 7.14 Cooperation and integration of European Standards Organizations (CEN, CENELEC, and ETSI) with global international partners

7.2.6 International Cooperation and Global Integration of ESOs

Figure 7.14 shows cooperation and integration of National Standards Bodies (NSBs) with the other interested industry partners (private companies) in European standardization bodies (CEN, CENELEC, and ETSI) and furthermore with the global international partners (ISO, IEC, ITU-T, ITU-R).

CEN and ISO are connected with the so-called Vienna Agreement (VA) from 1991 that allows simultaneous adoption of the new CEN and ISO Standard. VA also enables influence of ISO members to the contents of CEN standards and vice versa, with mutual respect of all political dimensions. As mentioned in Sect. 7.2.1, more than 30% of all CEN ENs are also ISO Standards. The double EN/ISO Standards have double role of automatic and identical implementation in 34 CEN Member Countries, as well as the global applicability. Except for the EN/ISO Standards, a relatively high number of ENs from CEN are closely connected to ISO Standards.

"Dresden Agreement, DA" connects in a similar way the two organizations CENELEC and IEC from 1996. In 2016, the "Frankfurt Agreement, FA" was signed with the reinforced primacy of electrotechnical standardization at the international level in the IEC. This agreement was necessary for avoiding the duplication efforts (approximately 80% of all European electrotechnical standards were identical to or based on IEC International Standards).

ETSI and the telecommunications part of the international ITU (ITU-T) are interrelated by the Memorandum of Understanding: "MoU in telecom sector." The radiocommunications part of international ITU (ITU-R) and ETSI are interrelated by "Agreement in Radiocommunications Sector." In 2000 and in 2012, ETSI and ITU renewed the MoU, which now encompasses the whole ITU.

Bibliography

1. European Union Intellectual Property Organisation (EUIPO), Internet page: euipo.europa.eu/ohimportal/. Retrieved 1.7.2019
2. Directive 98/34/EC, Internet page: eur-lex.europa.eu/LexUriServ/LexUriServ.do?uri=CONSLEG:1998L0034:20070101:EN:PDF. Retrieved 3.6.2019
3. CEN, Internet page: www.cen.eu. Retrieved 11.6.2019
4. CENELEC, Internet page: www.cenelec.eu. Retrieved 12.6.2019
5. ETSI, Internet page: www.etsi.org. Retrieved 6.6.2019
6. CEN-CENELEC Management Centre (CCMC), Internet page: www.cencenelec.eu/aboutus/MgtCentre/Pages/default.aspx. Retrieved 10.7.2019
7. Institute for Standardisation of Bosnia and Herzegovina, BAS, Internet page: www.bas.gov.ba. Retrieved 11.7.2019
8. General Directorate of Standardization-Albania, DPS, Internet page: www.zawya.com/mena/en/company/Albania_General_Directorate_of_Standardization-12619469. Retrieved 12.7.2019
9. Institute for Standardization of Montenegro, ISME, Internet page: www.isme.me. Retrieved 10.7.2019
10. The Standardization Association of the European Associations of Aerospace Industries, ASD-STAN, Internet page: www.asd-europe.org/aerospace-and-defence-industries-association-of-europe. Retrieved 11.7.2019
11. European Association for the Co-ordination of Consumer Representation in Standardization, ANEC, Internet page: www.anec.eu. Retrieved 12.7.2019
12. European Trade Union confederation, ETUC, Internet page: www.etuc.org. Retrieved 14.6.2019
13. European Trade Union Institute, ETUI, Internet page: www.etui.org. Retrieved 15.5.2019
14. European Trade Association for the Fire Safety and Security Industry, EURALARM, Internet page: www.euralarm.org. Retrieved 16.4.2019
15. European construction Industry Federation, FIEC, Internet page: www.fiec.eu. Retrieved 17.5.2019
16. Europe's Technology Industries, Internet page: ORGALIM, www.orgalim.eu. Retrieved 16.5.2019
17. Alliance of European medical technology industry association, MedTechEurope, Internet page: www.medtecheurope.org. Retrieved 15.5.2019
18. Small Business Standards, SBS, Internet page: www.sbs-sme.eu. Retrieved 12.5.2019
19. European Automobile Manufacturers` Association, ACEA, Internet page: www.acea.be. Retrieved 11.5.2019
20. European Cement Association, CEMBUREAU, Internet page: cembureau.eu. Retrieved 19.5.2019
21. European Disability Forum, Internet page: www.edf-feph.org. Retrieved 10.5.2019
22. European Emergency Number Association, EENA, Internet page: eena.org. Retrieved 11.5.2019
23. European e-invoicing Service Providers Association, EESPA, Internet page: eespa.eu. Retrieved 9.5.2019
24. European Federation for Elevator Small and Medium-sized Enterprises, EFESME, Internet page: www.efesme.org. Retrieved 2.5.2019
25. European Heating Industry, EHI, Internet page: www.ehi.eu. Retrieved 5.5.2019
26. European Partnership for Energy and the Environment, EPEE, Internet page: www.epeeglobal.org. Retrieved 6.5.2019
27. European Security Systems Association, ESSA, Internet page: essa.world. Retrieved 3.5.2019
28. European Solar Thermal Industry Federation, ESTIF, Internet page: solarheateurope.eu/welcome-to-solar-heat-europe/. Retrieved 15.6.2019
29. European eSkills Association, EeSA, Internet page: eskillsassociation.eu. Retrieved 16.6.2019
30. Near Field Communications Forum, NFC Forum, Internet page: nfc-forum.org. Retrieved 17.5.2019

31. Nanotechnology Industries Association, NIA, Internet page: www.nanotechia.org. Retrieved 2.5.2019
32. European Security in Health Data Exchange, SHIELD, Internet page: project-shield.eu. Retrieved 15.4.2019
33. International Commission on Illumination, CIE, Internet page: www.cie.co.at. Retrieved 16.6.2019
34. European co-operation for Accreditation, EA, Internet page: european-accreditation.org. Retrieved 14.5.2019
35. European Cooperation for Space Standardization, ECSS, Internet page: ecss.nl. Retrieved 12.5.2019
36. European Network of Transmission System Operators for Electricity, ENTSO-E, Internet page: www.entsoe.eu. Retrieved 3.5.2019
37. European Network of Transmission System Operators for Gas, ENTSOG, Internet page: entsog.eu. Retrieved 6.5.2019
38. European Association of National Metrology Institutes, EURAMET, Internet page: www.euramet.org. Retrieved 12.5.2019
39. European Organisation for Civil Aviation Equipment, EUROCAE, Internet page: eurocae.net. Retrieved 9.5.2019
40. International Federation for Structural Concrete, FIB, Internet page: www.fib-international.org. Retrieved 10.5.2019
41. International Federation of Standards Users, IFAN, Internet page: www.ifan.org. Retrieved 15.7.2019
42. International Telecommunication Union, ITU, Internet page: www.itu.int. Retrieved 12.6.2019
43. NATO Standardization Office, NSO, Internet page: www.nato.int. Retrieved 11.6.2019
44. International Organization of Legal Metrology, OIML, Internet page: www.oiml.org/en. Retrieved 10.5.2019
45. Universal Postal Union, UPU, Internet page: www.upu.int/en.html. Retrieved 11.5.2019
46. ZigBee Alliance, Internet page: www.zigbee.org. Retrieved 13.5.2019
47. Regulation (EU) No 1025/2012 of the European Parliament and of the Council of 25 October 2012 on European standardisation, amending Council Directives 89/686/EEC and 93/15/EEC and Directives 94/9/EC, 94/25/EC, 95/16/EC, 97/23/EC, 98/34/EC, 2004/22/EC, 2007/23/EC, 2009/23/EC and 2009/105/EC of the European Parliament and of the Council and repealing Council Decision 87/95/EEC and Decision No 1673/2006/EC of the European Parliament and of the Council. OJ L 316, 14.11.2012, p. 12–33
48. CENELEC Guide 8, The Vilamoura notification procedure for new national work and for the revision of national standards, Internet page: ftp://ftp.cencenelec.eu/CENELEC/Guides/CLC/8_CENELECGuide8.pdf. Retrieved 17.5.2019
49. Directive 2009/125/EC of the European Parliament and of the Council of 21 October 2009 establishing a framework for the setting of ecodesign requirements for energy-related products, OJ L 285, 31.10.2009, p. 10–35
50. Directive 2012/19/EU of the European Parliament and of the Council of 4 July 2012 on waste electrical and electronic equipment (WEEE), OJ L 197, 24.7.2012, p. 38–71
51. Official Journal of the European Communities, OJEC, Internet page: www.ojec.com. Retrieved 14.6.2019
52. Directive 2014/30/EU of the European Parliament and of the Council of 26 February 2014 on the harmonisation of the laws of the Member States relating to electromagnetic compatibility (recast), OJ L 96, 29.3.2014, p. 79–106
53. Associação Brasileira de Normas Técnicas, ABNT, Internet page: www.abnt.org.br. Retrieved 15.5.2019
54. American National Standards Institute, ANSI, Internet page: www.ansi.org. Retrieved 14.6.2019
55. Consejo de Armonizacion de Normas Electrotecnicas de las Naciones en las Americas (Span.), Council for Harmonization of Electrotechnical Standards of the Nations in the Americas (eng), CANENA, Internet page: www.canena.org. Retrieved 18.7.2019
56. Japanese Industrial Standards Committee, JISC, Internet page: www.jisc.go.jp/eng/. Retrieved 19.8.2019
57. KEYMARK, Internet page: keymark.cen.eu. Retrieved 21.8.2019

Chapter 8
National Innovation and Standards Circles

8.1 National Innovation Circle

The national innovation ecosystem or National Innovation Circle, NIC (shown in Fig. 8.1), consists of the same stakeholders as in Sects. 3.1, 6.1, and 7.1, but on the national level:

- Research and Education Institutions (REI), or Industry (IND) as a nurturing place of inventors
- Governmental and private organizations funding (FUN), forming the entrepreneurial level
- Marketing Organizations (MO), as organized places of marketers
- USers (US) and Users Organizations (UO) of a nongovernmental type

Globalization makes inventors, especially those coming from the industry (IND), to participate in the standardization process. In this case, the National Innovation Circle is shown in Fig. 8.1. As in the previous chapter, the stakeholder National Standardization Organization (NSO) was added. The five stakeholders are:

- Research and Education Institutions (REI), or Industry (IND) as a nurturing place of inventors, providing Funding (FUN)
- National Standardization Organization or National Standards Body (NSO or NSB)
- Industry (IND) producing invention
- Marketing Organizations (MO), as organized places of marketers
- USers (US) and Users Organizations (UO) of a nongovernmental type

However, the number of stakeholders changes in the case of inventors being holders of Intellectual Property Right (IPR). Then, the stakeholders are, as shown in Fig. 8.2:

- Research and Education Institutions (REI), or Industry (IND) (especially small and medium enterprises) as a nurturing place of inventors
- IPR organization, e.g., National Intellectual Property Organization (NIPO)

D. Šimunić, I. Pavić, *Standards and Innovations in Information Technology and Communications*, https://doi.org/10.1007/978-3-030-44417-4_8

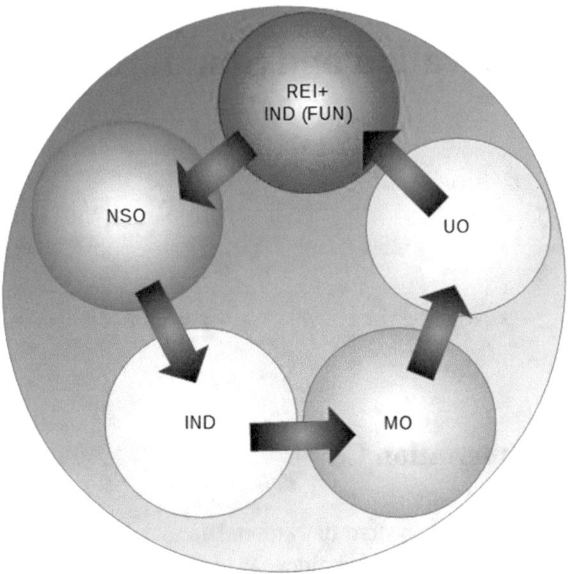

Fig. 8.1 NIC with the five stakeholders

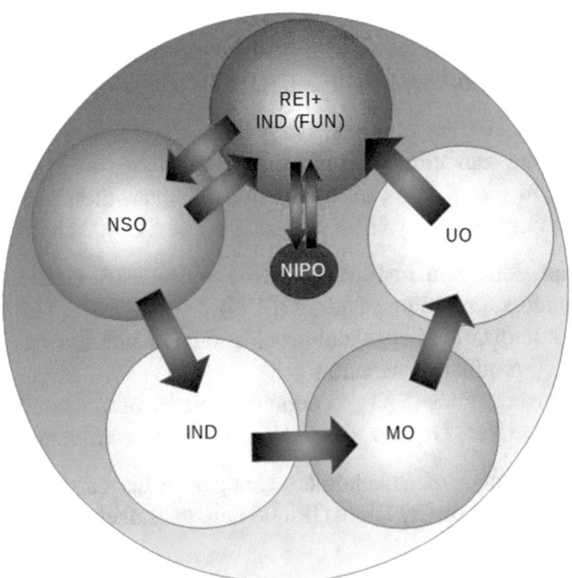

Fig. 8.2 NIC with the six stakeholders and NIPO

- National Standardization Organizations (NSOs)
- Industry (IND) producing invention
- Marketing Organizations (MO), as organized places of marketers
- USers (US) and Users Organizations (UO) of a nongovernmental type

The NIC has six stakeholders. Usually, it is a small and medium enterprise (SME) that is a REI with an IPR. In this case, REI first makes the invention and patents it with the patent organization (e.g., national intellectual property organization, NIPO). After obtaining the Intellectual Property Rights, REI promotes its invention in the NSO and contributes to the new product. Sometimes, the NSO buys the rights from REI or a company does it. The rest of the NIC is the same: after standardization, the industry produces the product, marketers market it, and users buy and use it. Users will give their opinion to REIs and industry how to improve or change it, and the circle is closed.

Quite often it happens that the inventor applies and gets the Intellectual Property Right (IPR) on the supranational level (so-called SIPO) (Fig. 8.3). An example of the supranational level is the European Union. In this case, the inventor applies to the level of the European Union, related to the IPR. If the IPR relates to the patent, the inventor (which could also be an SME or Industry as IPR Applicant) applies to the European Patent Office (EPO) [1]. EPO is the international supranational organization with 38 member states. If the IPR is a copyright, the inventor contacts the European Union Intellectual Property Office (EUIPO) [2] that provides access to creative content protected by copyright. This is discussed in length in Chap. 4.

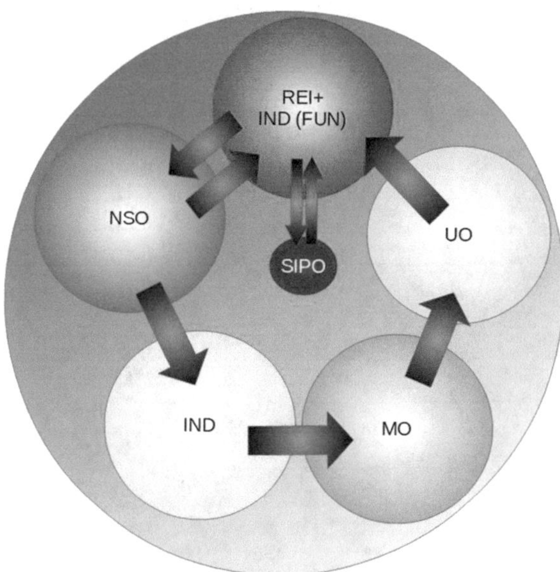

Fig. 8.3 National Innovation Circle with the six stakeholders and supranational IPR organization (SIPO)

In this simple innovation circle, there are several possibilities. One is that the inventor holds IPR from NIPO and decides to invest this knowledge to the Supranational Standardization Organization (SSO). This is shown in Fig. 8.4.

Another possibility is that the inventor holds IPR from SIPO and decides to input this knowledge to the National Standardization Organization (NSO) or to the Supranational Standardization Organization (SSO). This is shown in Fig. 8.5.

An example of a Supranational Standardization Organization is one of the European Standards Organization (ESOs) of the European Union (CEN, CENELEC, or ETSI), and an example of the supranational IPR organization is the European Patent Office (EPO) of the European Union.

Thus, we can consider various configuration types of the National Innovation Circle. Despite the many possibilities, only the four most important cases are given in Table 8.1.

The abbreviations related to the configuration types are the same as in Fig. 8.2, i.e., REI stands for Research and Education Institutions, FUN stands for Governmental and private organizations funding, MO stands for Marketing Organizations, UO stands for Users Organizations of a nongovernmental type, SO stands for Standard Organization, and IPO stands for Intellectual Property Organization. Furthermore, "GL" means "Global Level," "SNL" means "SupraNational Level," and "NL" means "National Level."

As shown in Table 8.1 and Fig. 8.6, a Globally Oriented NIC (GONIC) has global, supranational, and local characteristics. Knowledge is contained in the global community (on the Global Level, GL), funding is of the national type, the

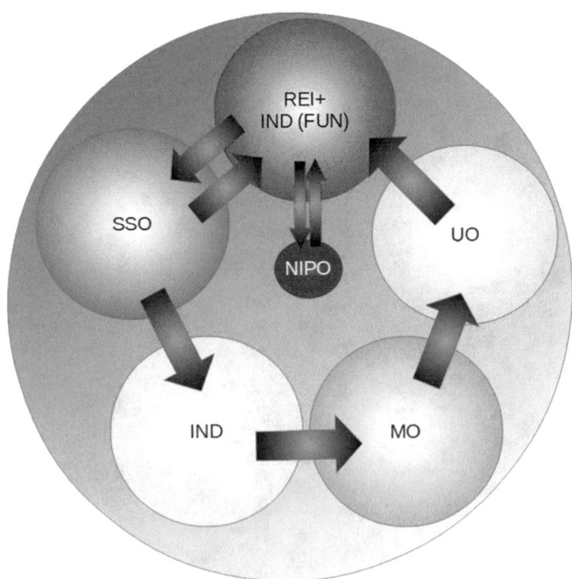

Fig. 8.4 NIC with the National IPR Organization (NIPO) and Supranational Standardization Organization (SSO)

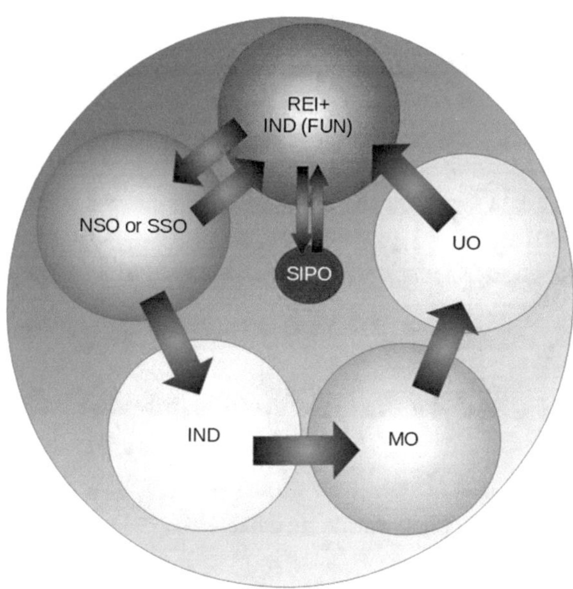

Fig. 8.5 National Innovation Circle (NIC) with the Supranational IPR Organization (SIPO) and National (NSO) or Supranational Standardization Organization (SSO)

Table 8.1 The four important configuration types of National Innovation Circle

NIC type Stakeholder	GONIC	LMNIC	LKNIC	LNIC
REI	GL	SNL	NL	NL
FUN	NL	SNL	SNL	NL
MO	GL	NL	GL	NL
UO	GL	NL	GL	NL
SO	GL	GL	SNL	NL
IPO	GL	SNL	SNL	NL

market is oriented toward price competition, and the nature of users is global, together with the standard organization and Intellectual Property Organizations. The second type, Local Market NIC (LMNIC), has characteristics of supranational knowledge and funding, but the market and users retain a local nature (i.e., on the National Level, NL). The standardization organization is global, and the Intellectual Property Organization is supranational. The third type, Local Knowledge NIC (LKNIC), has a market and users with global nature; funding, standardization, and intellectual properties of the supranational; and knowledge of national nature.

Finally, the fourth type, Local NIC (LNIC), is characterized by all characteristics of the local nature.

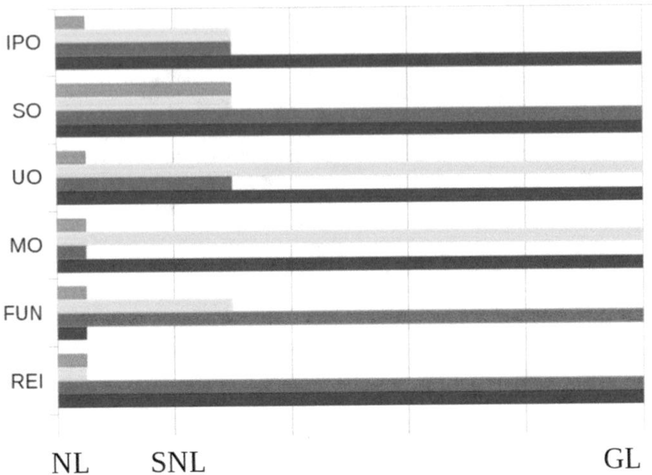

Fig. 8.6 National Innovation Circle with the six stakeholders

8.2 National Standards Circle

The National Standards Circle (NSC) consists of the four main stakeholders:

- National Standards Body
- Authority
- Industry
- Societal stakeholders

The National Standards Body (NSB) has a central position in the standardization stakeholders circle. The other three stakeholders are already explained. However, on the national level, the authority is a national government. Industry as a stakeholder represents national industry. Finally, societal stakeholders are the same as on supra-national level. Therefore, this part of the chapter is devoted only to the NSB.

In most cases, NSB is established as a specialized governmental agency, but it can be an entity already existing in a national government. An example may be the Ministry of Foreign Affairs, or in the case of, e.g., telecommunications, it could be a Ministry of Communications. In certain cases, it may even be a private entity.

NSB provides many services to the standardization activities on national, regional, and international levels. Figure 8.7 shows examples of NSB activities on the international level, authorization and management of national delegations, including actions related to coordination and/or strategy of standardization activities on the national level for the international community. NSBs also help preparations and sometimes host the international meetings. NSBs develop national strategies and policies for standardization in all fields, including ICT.

Figure 8.7 summarizes NSB activities as follows:

- NSB authorizes participation on international level, including the private sector

Fig. 8.7 Activities of NSB

- NSB provides bidirectional centralized reporting to and between the national and the international community
- NSB represents its country, one of the Members

On the national level, the NSB coordinates activities in the private and the public sector. This is important on all the levels, including the regional and international ones. Its importance is in giving accreditation to the correct experts, i.e., delegates who can represent the country in the required settings (regional, international). The NSB is a leader and a head for the national delegation on the international meetings. Per itself, this avoids any kind of confusing or conflicting situations across different but related regional and international partners. It gives an improved operation of delegation management policy before, during, and after an international meeting. The result is the increase of the awareness and interface for exchange of information on the national level between the NSB and involved experts. Finally, NSB can inform the leading experts on the national level about the results and of the newest information on the international level and vice versa (i.e., reporting of the national activities to the international level). This means, among all the other benefits, reducing costs for all the national stakeholders.

Figure 8.8. summarizes activities of NSB on the national level:

- Management of a national secretariat
- Dissemination of information from international standardization organizations to relevant stakeholders
- Development of national policies related to standardization
- Coordination of capacity building for international standardization activities
- A centralized place for all formal standards on the national level in the given area

The National Standardization Secretariat (NSS) has the leading role in the preparation processes for international meetings. This includes discussions on the national level, leading to the developed joint national position on required issues. NSS authorizes national delegation/experts for participation in international meetings.

Fig. 8.8 NSB activities on the national level

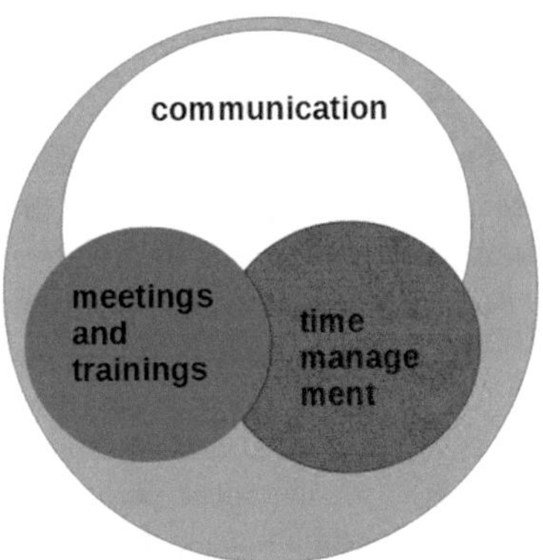

Fig. 8.9 Administrative functions of the NSB

Since it represents its own country before, during, and after international meeting, the NSS also develops a management policy for it. After the meetings, the NSS reports both to the international community and national stakeholders. Finally, the NSB is the central place for formal standards in the country. Of course, this is only for the area that NSB is declared and, thus, competent.

The NSB performs also administrative functions. The three most important (shown in Fig. 8.9) are:

- Communication
- Time management
- Meetings and trainings

Communication functions encompass hosting and maintaining the web site, all e-mail lists, and all the records. Under this function is also monitoring of all kinds of communications between regional and international organizations, in which they are the members, representing its country and its national distribution.

Time management functions of the NSB are related to timely payments of all the annual dues to all the regional and international organizations, in which the NSB is the member, representing its country. Time management functions are also related to timely required responses and contributions from the related national committee chairs or experts to all the regional and international organizations, in which the NSB is the member and represents its respective country.

Functions related to the meetings arrangements are an assistance to the national standardization secretariat committees for the physical arrangements, e.g., meeting rooms and announcement and document distribution. The training arrangement of the national delegates is another part of this administrative function.

National Standards Bodies can have three kinds of general structure, dependent on the capacity level and on the importance of the standardization organization on the national level. These three levels are (Fig. 8.10):

- Passive
- Half-active
- Active

The passive NSB has a high general interest in the participation. However, the organization is based on the principle of a minimum energy. In reality, the passive NSB has almost no capacity or a very small and limited capacity for a minimum involvement in any of the working areas of the regional and international organizations, in which it participates as the member and represents the community of the country.

A half-active NSB is involved in some activities of all the regional and international organizations, in which it is the member and represents the country.

Finally, the active NSB actively collaborates in almost all or in all the activities of all the regional and international organizations, in which it is the member and represents the country.

For all three levels of the NSB development or for all three NSB's configurations, the crucial part of the NSB is a National Advisory Committee (NAC). NAC is created for all the regional/international organizations in which the NSB participates. NAC is open to private and public sector participants.

A passive NSB is defined with an agreement that an agency becomes National Standardization Agency (NSA). NSA handles (Fig. 8.11):

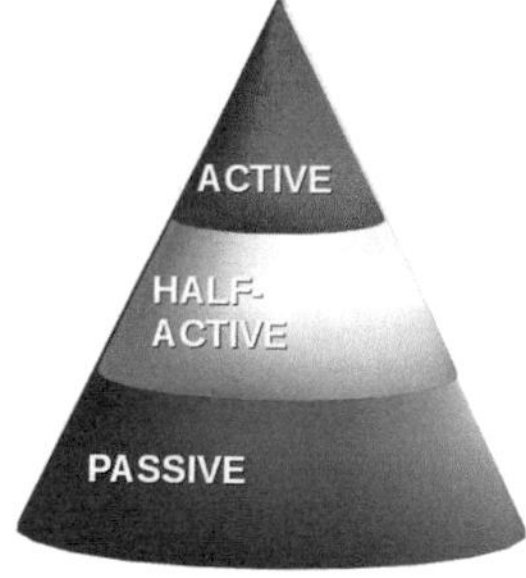

Fig. 8.10 Three levels of development of NSB: passive, half-active, and active

Fig. 8.11 An organization of a passive NSB

- Establishment of the national processes and procedures
- NSB approval of contributions from the country to all regional and international standardization meetings that the NSB represents on the national level
- Official communication between the regional and the international standardization meetings that the NSB represents on the national level to the NSB's approval on behalf of the Member Country
- Approval of the private sector applications to become Sector Members, Associates, and Academia participants
- Appointment of the Chair of NAC for all the regional and international standardization meetings that the NSB represents on the national level

For a passive NSB, function of NAC is threefold (shown in Fig. 8.11):

- Manage preparatory actions for all the regional and international standardization meetings that the NSB represents on the national level.
- Propose policies related to the participation at all the regional and international standardization meetings (in which the NSB represents the country) to the NSB approval.
- Provide the representation that monitors activities, participates in the meetings, and responds to enquiries for all the regional and international standardization meetings to NSB approval.

NAC can form ad hoc groups that are open for private and public participation. NAC also appoints Chair and Terms of Reference, based on the need. Tasks and responsibilities of ad hoc groups on the national level include development of national positions on topics that are not associated with a meeting. Related to the international level, ad hoc groups prepare contributions and report about the results of the meetings.

The fourth entity of the simplest, passive NSB is the National Standardization Secretariat (NSS). NSS is either a part of NSA or contracted by the NSA. Its activities are in the national distribution of all relevant documents (information, announcements, meeting documents), storage of the record, development and maintenance of the NSB (and NSA) web site, and sending all the required and the relevant information to the international standardization organizations in whose activities the NSB acts as a member.

The half-active NSB is characterized by an establishment of a clearly designated NSA that handles the stable and permanent funding that includes a permanent secretariat (NSS). Half-active NSA handles:

- Establishment of national structure, i.e., of permanent national groups with active involvement and contribution in at least one of the working groups of international standardization organizations
- NSB approval of contributions from the country at all the regional and international standardization meetings that the NSB represents on the national level
- Official communication between the regional and international standardization meetings that NSB represents on the national level to the NSB approval on behalf of the Member Country
- Approval of the private sector applications to become Sector Members, Associates, and Academia participants
- Appointment of the Chair of NAC of all the regional and international standardization organizations that the NSB represents on the national level
- Appointment of the Chairs of NAC for at least one of the regional and international standardization organizations (denoted as X-NAC) that the NSB represents on the national level to NSB approval

For a half-active NSB, the function of the NAC is threefold (shown in Fig. 8.12):

- It manages preparatory actions for all the regional and international standardization meetings that the NSB represents on the national level.
- It proposes policies related to the participation at all the regional and international standardization meetings that the NSB represents on the national level to the NSB approval.
- It provides a representation to all the regional and international standardization meetings that the NSB represents on the national level via separate sub-NACs (denoted as X-NAC) for separate international standardization organizations, for which the NAC is a parent body.
- It maintains national procedures for all the meetings of the regional and international standardization organizations that the NSB represents on the national level.

The half-active NSB establishes separate NAC for some of the regional and international standardization meetings that the NSB represents on the national level.

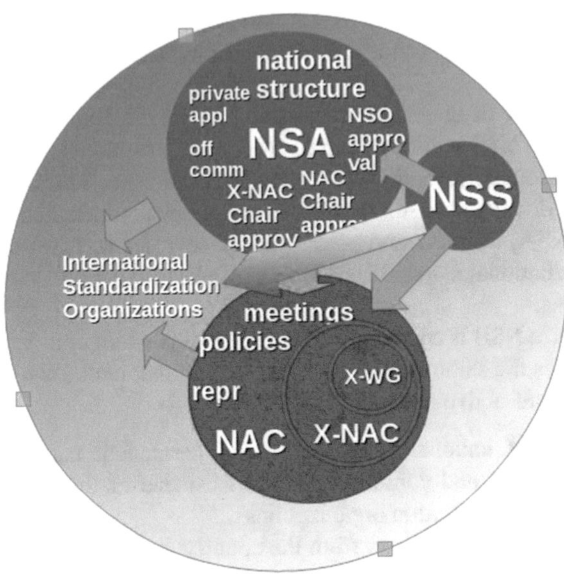

Fig. 8.12 Organization of half-active NSB

It can be called X-NAC, where X could mean ISO, IEC, and ITU (or CEN, CENELEC, and ETSI in Europe).

X-NAC is a parent body to the national Working Groups, denoted as X-WG (X depends on the corresponding international or regional standardization organization). Thus, all the X-WG Chairpersons are included in X-NAC. X-NAC is open to private and public participation. The Chair of X-NAC is appointed by the NSA. Responsibilities of the X-NAC include preparations for X organization meetings, policy proposals to X participation for the NSA approval, provision of a representation, and creation of relevant subcommittee(s) of X-WG.

X-WG is also open to private and public participation. Most often, it has the same title as the WG in the X-organization with the Chairperson appointed by the X-NAC Chair. Its responsibility is to make all the preparations for meetings of X-organization, to propose national delegation members, and to report back to national level about X-organization meetings and results.

The NSS is either within the NSA or it is contracted on a continuous basis by the NSA. Responsibilities are mostly oriented to the national distribution of relevant information and documentation, which includes developing and maintaining the web site and record keeping.

A NSB becomes an active NSB when the country participates at nearly all international standardization meetings with many activities in various WGs and SubGroups (SG) of different international standardization organizations (Fig. 8.13). For example, if these organizations are ISO, IEC, CEN, and CENELEC, then the NSB should have four NACs. This means that the NSA is quite well organized and that it is capable of addressing various policy issues. The process of change from a

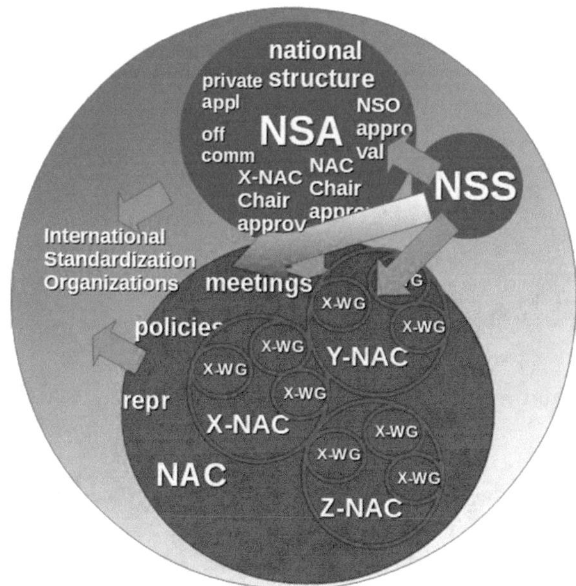

Fig. 8.13 Organization of an active NSB

half-active to the active NSB is usually slow, and the structure slowly expands by involvement of more and more experts in the standardization work. This means that the active NSB can cover more areas in general. An especial property is that the areas are even more specialized areas than in the previous case of the half-active NSB. This requires a stable secretariat (NSS) and a stable funding.

Of course, the goal of every country is to have an active NSB in place to serve the national standardization needs.

Bibliography

1. European Patent Office (EPO), www.epo.org. Retrieved 11.7.2019
2. European Union Intellectual Property Office (EUIPO), https://euipo.europa.eu/ohimportal/en. Retrieved 10.7.2019
3. A.L. Wirkierman, T. Ciarli M. Savona, *Varieties of European National Innovation Systems*, 13/2018 May, ISI Growth, working paper, H2020 RIA under grant No 649186
4. B. Godin, J.P. Lane, Pushes and Pulls: Hi(S)tory of the Demand Pull Model of Innovation. Sci. Technol. Hum. Val. (ST&HV) **38**, 621 (2013)
5. ISO Central Secretariat, *Fast Forward, National Standards Bodies in Developing Countries*, 2008, Internet page: https://www.unido.org/sites/default/files/2008-10/fast_forward_0.pdf. Retrieved 11.7.2019

Index

D. Šimunić, I. Pavić, *Standards and Innovations in Information Technology and Communications*, https://doi.org/10.1007/978-3-030-44417-4

Printed in the United States
by Baker & Taylor Publisher Services